Microcomputers
in Building Appraisal

Microcomputers in Building Appraisal

Peter S. Brandon MSc(Arch), ARICS, MACostE
with
R. Geoffrey Moore BSc, CEng, MICE

Nichols Publishing Company
New York

Nichols Publishing Company (W. G. Nichols, Inc.)
PO Box 96, New York, New York 10024 and
155 West 72nd Street, New York, New York 10023.

Library of Congress Cataloging in Publication Data

Brandon, P.S. (Peter S.)
   Microcomputers in building appraisal.

   Bibliography: p.
   1. Building — Estimates — Data processing. 2. Buildings
— Valuation — Data processing. 3. Microcomputers.
I. Moore, R. Geoffrey. II. Title.
TH437.B7 1983   333.33'8   82-22435
ISBN 0-89397-147-2 (pbk.)

First published in United States of America 1983 by Nichols Publishing
Company
First published in Great Britain 1983 by Granada Publishing Ltd
Frogmore, St Albans, Herts AL2 2NF
and 36 Golden Square, London W1R 4AH

Printed in Great Britain by Richard Clay (The Chaucer Press) Ltd,
Bungay, Suffolk

DISCLAIMER
The publishers and the authors disclaim all responsibility for use of the programs included in this book for any purposes whatsoever. They are general
rather than specific examples only. It is therefore the responsibility of the reader
to check the suitability of the program and the calculations, the technique
and the methodology adopted for the job to which he is applying it before
implementing the results.

Contents

Dedication

To: Mary   Audrey
     Samantha Andrew
     Juliette   Joanna
     Robin

Information on
'starter pack' programs

The starter packs at the end of this book (section III) are available for the CBM, PET and the Apple II on disk or cassette.

The price of each disk or cassette, on which all the programs are recorded, is £40.00 (US$80) including postage and packaging.

Please order from:

C.O.E.D.
Computer Orientated Educational Developments
8 Huxley Close
Locks Heath
Southampton
SO3 6RR
Great Britain

enclosing a postal order or cheque for the full amount. Remittances should be payable to C.O.E.D. Please state whether disk or cassette is required and on which machine the programs will be used.

Preface

The aim of this book is to introduce those members of the construction industry with little or no knowledge of computing to the use and programming of microcomputers.

The need for such a book became evident from conversations held with delegates on short courses for building professionals run as an introduction to microcomputer applications. To understand the applications it is first necessary to have some knowledge of the machine and the way it is programmed. There appeared to be a demand for a publication which presented the facts in an easily understood manner and, more importantly, introduced the subject through the medium of the building problem. At the same time it was also evident that students of construction, architecture and quantity surveying did not have a book related to their practical computer requirements.

For both practitioner and student there appeared to be three basic needs. The first was for some underlying theory which gave a background to how a computer worked, what its limitations were, and how problems should be structured for the machine. It was stressed by those consulted that this should be an appreciation of these factors rather than a detailed analysis and that it should be related to practice and free from technical jargon where possible. Section I gives an introduction to this kind of information, although the authors are fully aware that they have merely scratched the surface of a very complex technical subject and it has not always been possible to avoid jargon. However, hopefully, most terms have been clearly defined.

Secondly there was a need for an introduction to programming using simplified examples of problems found in the building industry to improve the relevance and understanding of the technique. Section II gives such an introduction for the language of BASIC, in readiness for the 'starter pack' programs (section III). However this is not an attempt to compete with the many excellent general texts which are available and which are often able to cover the subject in greater depth. Despite this proviso it should be possible for the reader to sit at his computer and learn the rudiments of programming by trying out the routines given.

Thirdly there was a need for some example routines that would give practitioners a good start in developing their own programs. Therefore the final section gives some longer programs which apply the knowledge gained and should provide the basis of suitable office applications. Each professional will no doubt want to develop these programs in his own way and for this reason only the bare skeleton of the programs ('starter packs' as they have been termed) is provided, allowing the reader to add his own routines.

Although the subject matter of the book is geared towards microcomputers, the programs themselves can be used on any computer which allows the use of the BASIC language. Owing to the different dialects of the BASIC language there will no doubt be slight alterations required for manufacturers other than the Commodore PET, Apple and Tandy microcomputers for which the instructions are given. Reference should be made to the manufacturer's reference manual where there is doubt. An attempt has been made to keep to the most common BASIC commands wherever possible.

While every effort has been made to ensure the accuracy of the programs an occasional 'bug' may be discovered. The authors would be grateful if these could be reported to them for correction in future editions.

Lastly we would like to stress that we do not believe that every building professional should become a programmer. What we do believe is that a knowledge of the computer even at the introductory level of this book will be helpful in understanding the potential of the machine; the manner in which problems need to be structured; the instructions required by a systems analyst and/or programmer; and the problems of introducing the machine into the office.

The text is essentially a practical one and to get full benefit it will be necessary to try the programs out on a machine and enhance them with your own ideas. No amount of reading will compensate for 'hands-on' experience. May we wish you well.

Acknowledgements

In a book of this nature, a debt is owed to a large number of individuals who directly or indirectly have contributed by giving their time and advice. They cannot be held responsible for any defects but they can take credit for shaping the book in its current form. We would thank all who have been consulted but in particular the following:

Phillip Main, the programmer at the Department of Surveying, Portsmouth Polytechnic, Portsmouth, England, who took much of the basic raw program material and transformed it into a presentable form for publication, and in addition contributed much original work of his own.

Bert Coates, who supplied the cartoons which have added a touch of humour to the text. In addition his help and advice on the content has been invaluable.

Michael Powell, Head of the Computer Department at Portsmouth Polytechnic, whose comments on the early chapters were much appreciated.

Pat Curtis, Eileen Dobson, Audrey Morford and Ruth Hawkesworth of the Department of Surveying at Portsmouth Polytechnic, who undertook much of the typing, sorting and draughtsmanship. To these must go a special thanks for their patience, understanding and skill.

Lastly we would like to thank our families who have forgone a good deal of normal family life in order to bring this book into being. To these and many more we acknowledge our debt.

Extracts from BS 4058: 1973 are reproduced by permission of the British Standards Institution, 2 Park Street, London, W1A 2BS, from whom complete copies of the standard can be obtained.

SECTION I
The Machine in Office Practice

CHAPTER 1
Introduction

Imagine yourself seated in your office surrounded by the essential paraphernalia of the building professional. There will be shelves full of reference information, called upon when the inadequacies of human memory require some support. An 'In' tray and 'Out' tray on the desk will signify the need for human decision making and filing cabinets against the wall will hold records of those decisions. In one corner there may be a drawing board or work area where human ideas are rather laboriously transferred from the mind to a more tangible medium for transmitting those ideas to others. Lying on the desk are the tools of trade such as the stapler, typewriter, paper punch and possibly the calculator, together with the telephone to contact the outside world.

The centre of this office is you, the human being, around which the gadgets, the information and the furniture are designed. Using your senses, probably you will have been for many years the screener, manipulator and processor of the information which passes through the office. In addition you will very often be the creator of new information which someone else will need to process.

The interesting aspect of this scene is the size of room required for you to work (not merely enjoy your surroundings), the cumbersome nature of the materials used and the slow speed at which communication and manipulation of information takes place. Until recently this was unavoidable, because there was no satisfactory economic alternative to paper and brain. The tools that were developed had replaced or enhanced human limbs, but rapid enhancement of the capability of the brain was not available. In the natural course of events refinement of the brain has of course taken place throughout its formal and informal education. It has built up its own procedures or algorithms including sorting, screening and even routines such as the learning of multiplication tables, to cope in a more efficient manner with the demands placed upon it. With the introduction of the computer at least part of the function of the brain is being enhanced. This is the part that deals with the routine calculation and processing of information. In addition office information can be held on the microcomputer in a much more compact form, which can be accessed, in most cases, more efficiently.

The traditional tools of trade and the appearance of the office are changing fast and in some cases the conventional apparatus is becoming obsolete. However we still have a long way to go before we reach the utopian offices of the future as presented in the glossy magazines. Now that primary schoolchildren are beginning to use computers as a natural tool, the pressure to change, arising from this education of our young people, will eventually overcome the prejudices of the older generation. Even since the late 1970s the amount of interest in computers, stimulated by the media and home computers, has increased enormously in the offices of the building professions.

Building appraisal

This book aims to introduce those members of the professions and students who have very little or no experience of computers to some of the basic concepts, knowledge and skills required to gain an appreciation of how to use the machine for building appraisal.

Building appraisal has been chosen as the medium for this introduction because it has the ingredients for harnessing the power of the computer. To appraise a building, no matter in what sense that appraisal is required, the principles are very similar. The problem must be identified and structured; a method of solving the problem must exist together with a suitable form of measurement. In many cases a representation of the problem in some other form, i.e. a model, must be constructed. In addition it is not uncommon for several alternatives to be considered.

These processes are ideally suited to computer execution. The machine needs, as we shall see in later chapters, a formal solution to the problem and it is particularly good at repetitive tasks, making it ideal to evaluate alternatives where the principles of evaluation are the same.

Building appraisal is also a good vehicle for computing as it crosses the boundaries between the established building professions. One impact of microcomputers will be that one man will be able to cover a wider range of skill. It may be, therefore, that the specialisms which have developed over the years, usually stemming from the main discipline of architecture, will begin to be reduced in number. Already architects working on a computer-aided design system can alter specifications and building geometry and get an immediate cost appraisal. Whether in the long term the machine will be able to judge the consequences of such a decision as well as an expert cost adviser it is difficult to say. Whether architecture will be the only building profession in the long term future (together with the contractor), with the other currently recognised professions providing merely technical back up, is a matter for conjecture. What can be said is that the present entrenched and apparently static positions of the traditional professions will inevitably change.

Rise of the machine

So how did this revolution arise and why is it only comparatively recently that architects, quantity surveyors, engineers and contractors have in general taken on board the power of the machine? There have been of course a dedicated few who have patiently worked at the development of computers for building applications. In several cases their investment in the new technology is beginning to pay off; with others, disillusionment has set in and the many thousands of hours required for development of major systems has been wasted. Before considering this matter it is worth reviewing the historical perspective for the development of the machine.

History

To a modern readership it is probably unnecessary to spell out once again the history of computing from the desk calculators of Blaise Pascal and Baron von Leibniz, through Charles Babbage and William Burroughs, to the first IBM Automatic Sequence Controlled Calculator of Professor Aiken in 1937. We are now in what is sometimes called the sixth generation of the electronic digital computer. (Digital computers use pulses to set on/off electronic switches to work and therefore use discrete increments to handle numbers, whereas analogue computers use another medium, e.g. flow of electrical current, to represent a physical occurrence such as heat flow through a building. Analogues are therefore continuous or measuring as opposed to discrete or collating in their operation.)

The generations of computer systems can be summarised as follows:

Generation 1	Vacuum tubes
Generation 2	Transistors
Generation 3	Hybrid and integrated circuits
Generation 4	Medium and large scale integrated circuits
Generation 5	Minicomputers
Generation 6	Microcomputers

New generations are on the horizon and may soon be with us as developments are moving fast.

Impact on the user

The impact of these developments as far as the user of a system is concerned has been dramatic:

(i) cost has been reduced to in some cases less than 1% of its 1960s cost for comparable power;

(ii) physical size has been reduced so that we now have powerful machines able to sit on a desk top;

(iii) sensitivity to the environment in which the microcomputer has to work has been reduced to the same level as other office equipment;

(iv) the power of the machine has increased enormously in both speed and memory capacity;

(v) reliability has increased — in fact with microprocessors and the automatic testing procedures that are adopted it is difficult to make their operation unreliable. (Testimony to this fact can be found in the reliability of digital watches, which use a small microprocessor.) It is the mechanical devices such as disks and printers which tend to cause trouble if it arises;

(vi) backing store in the form of floppy disks has become inexpensive and of higher storage capacity.

It is not only in the hardware or machinery that the change has occurred. The software or programming of the machine has also undergone considerable changes, such as:

(i) the number of computer 'languages' has increased, each of which has been developed for a specific application;

(ii) some languages have become more like spoken English and therefore easier to use;

(iii) the operating system of the machine has become more sophisticated and less onerous to implement (sometimes to the point of not being evident to the user);

(iv) software development has now reached the point where the machine can undertake a certain amount of self-programming once a basic structure has been provided by the operator;

(v) manuals related to the operation of the machine have become simpler and in many cases include a self-teaching text.

Despite these advances in software it is generally recognised that software development has not kept up with the developments in hardware. It is still a very time consuming task requiring great skill to write an efficient program of reasonable length. Although the speed and capacity of the machinery will no doubt continue to improve for comparable cost, much more time and money needs to be spent on software to obtain a similar measure of advancement.

This lack of development in software is a major problem which is particularly relevant to the building industry and consequently the full potential of the machine is not being realised.

Response of the building industry
An industry which is as diverse as building, with such a large number of units of production, extensive technical data and complex managerial and procedural arrangements would appear to be an obvious target for computer applications. The data handling capabilities, the rapid number crunching facilities and the speed of output would seem to make the machine an essential tool for the building professions and builders alike.

No doubt these factors were in the mind of those firms that jumped on to the bandwagon in the late 1950s and early 1960s. Many of the large architectural practices, spurred on by the 'glossies' arriving from abroad; preparers of schedules, bills of quantities and specifications trying to overcome the shortages of technical staff; and contractors needing to handle the complexities of their payroll; all took an active interest in this new phenomenon. Engineers too, followed the enticing path to worship at the shrine of the unknown calculator and were excited by the mountain of paper thrown at them by the touch of a button.

By the middle of the 1970s only a few of these enthusiasts remained. Contractors had found that the machine could cope with their payroll and invoicing but the economy of the investment was still in question. Architects had found the machine constraining and inflexible while quantity surveyors discovered that an efficient manual sorting system such as 'cut and shuffle' was a better economic proposition than all the manual data preparation and error checking required by the machine. Engineers were perhaps the exception in that they had tailor made techniques using the application of formulae ideally suited for the machine.

It is worth considering the major reasons for this disillusionment in order that we do not make the same mistakes again.

Early problems
Perhaps the major contributory factor to the ensuing disappointment was the high expectations of those using the machines, sometimes owing to extravagant claims by the manufacturers, bureaux and software houses. The initial investment, which in those days was exceptionally expensive, very often running into a hundred thousand pounds or more, was based on the assumption that the machine could undertake current techniques much more efficiently than human effort.

The computer was expected to take on board a technique which had been developed to overcome the deficiencies of human brain power, movement and capacity, and immediately improve its performance. Instead of asking, 'How can I adapt my technique to suit the vast potential of the machine?', the professions asked, 'How can I apply the computer to my current technique?' In some cases, it paid off after many years by persistent effort. The initial investment in developing systems from scratch, purchasing the machine, providing a controlled environment and training staff was enormous, and many firms had to write off large sums of money and this has naturally made them cautious of any further involvement.

Other mainly psychological factors that contributed to that early disillusionment can be summarised as follows:

(i) Most early computer systems used 'batch' processing where the input data was placed on cards before being processed — usually in a room or office isolated from the people who prepared the data. The designer or cost adviser would need to wait until the machine had processed his information, output any errors, and then printed his results. The correction of errors very often required several reruns and this was obviously time consuming and frustrating.

(ii) The machine was geographically remote from the work force and therefore psychologically a barrier and mystique was created between the professional and the computer, which in many cases led to a lack of credibility when things went wrong. It was argued that by removing the processing from the practitioner there was little interference with his normal technique. Unfortunately many practitioners also felt little responsibility or involvement with the machine and consequently a sympathetic relationship between man and computer did not develop.

(iii) Decisions as to whether a computer was required or not were sometimes made by a partner or small group who then imposed that decision on the firm or organisation without selling the idea to those who were required to operate the new system. Grass roots reaction to the unknown is very often retrenchment to the status quo.

Inevitably in the long run it was the economic factors which had more impact and led to a final disengagement. The expense of the system as already outlined could not be justified unless there was a critical problem which could not be resolved by other means. Those firms that were preparing bills of quantities and schedules, and could not get skilled technical help, persisted, and, despite the problems of adapting manual techniques to machine operation, are now reaping the benefits. Those architectural practices that were involved in system building, and which did not always find it possible to attract the right personnel, found the machine could improve the speed at which drawings were produced and have now found themselves leaders in the field. However many of those who experimented did not have these significant problems and consequently were able to revert to traditional methods when the investment appeared to go sour.

The role of research

Fortunately during this period of disillusionment a number of educational institutions saw the potential of the machine and began to develop techniques for the benefit of the building industry. ABACUS at the University of Strathclyde, Glasgow, Scotland, and the Computer Applications Group at Bristol University, Bristol, England, are examples within schools of architecture who saw a long term benefit for the design team and explored the possibilities. Although a large number of programs of considerable interest were developed, it is true to say that for a number of reasons only a few of them have been adopted in practice. However, they have left pointers for the future which will be capitalised upon by the professions when the time is right.

For building contractors and bills of quantity applications the research has largely been undertaken by individual firms or groups of practices or organisations such as local authority consortia or central government agencies. In the case of local government departments the fact that a machine had already been installed for other purposes (usually for use by the treasurer's department) meant that overheads for development were considerably reduced. It was therefore almost obligatory to use the machine wherever possible. Even so there are many stories of woe to be told with regard to implementation even in these institutions. Some development work for contractors has been undertaken in the centres of education, such as the estimating systems developed at Loughborough University, Loughborough, England, and the cost models developed in some of the UK polytechnics.

For those who wish to gauge in which direction industry will move in the future it is an interesting exercise to visit these educational institutions. One subject which is being given particular attention at all levels of the computing spectrum is the organisation of construction data bases. It has been realised that most appraisal techniques are only of use if the data is available to operate them. It has also been realised that considerable mileage can be achieved in economic terms if the data on technical matters and

project information held in the office can be accessed and reproduced at speed. The rapid increase in the number of word processors
in offices is evidence of this interest. The next step will be the
manipulation of that information in models, budgeting and appraisal techniques. A move in this direction has already begun and most
offices now have a microcomputer tucked away in the corner. At
the moment many of these micros have severe restrictions on memory
capacity and speed of operation. As time goes on there is little doubt
that these constraints will be removed and large systems will be able
to be developed which will allow access to a comprehensive data
base.

The microcomputer

So what has changed that gives us confidence in the future of the
machine? We have already outlined the very great advances in hardware technology and this has resulted not only in a much more
attractive economic investment but has also overcome many of the
psychological factors which were contributory factors to the early
disillusionment.

The small size and reduced cost of the machine means that it
really is a feasible desk top tool. No doubt in time it will become an
instrument the size of a small calculator. Already there are rumours
of pocket machines which will be able to hold vast libraries of
information. At the present time a major obstacle to a really
dramatic development in the flow and exchange of information is
the lack of compatability between machines. Very few of the smaller
microcomputers are able to use each other's programs and data
without rekeying of the instructions. With the advent of the CP/M
(control program/monitor) operating system, fast becoming an
industry standard, and a more standard form of computer language,
in time even this constraint will disappear.

Small independent work stations attached to a large mini or main
frame computer are already available in the middle range of microcomputers and this is likely to be the pattern for office use in the
immediate future. The mini can be used to its maximum advantage
by queuing the data arriving from the work stations, and meanwhile
the inputting of data and the independent use of the microcomputer
are not affected. As the micro becomes more powerful then the need
for the larger back up storage or the increased speed of the mini may
no longer be a necessity.

Most microcomputers are interactive; by this we mean that the
program can be written to stop at various places for such things as
keying in data, answering questions and outputting interim results

upon which a decision must be made. In this sense a 'conversation' can be held with the machine, although it should be stated that the communication will be rather stereotyped! The problem of batch processing, where data is prepared remote from the machine, is therefore no longer a factor as the operator can exercise more control over the program and can identify with the processing being undertaken. The remoteness of the machine has been overcome and the professional feels a sense of independence from the computer personnel who tended to rule this part of his life in days gone by. By his very nature as an independent adviser to his client the building professional wishes to maintain his independence and control of events. The largest complaint heard of computers in professional offices in the 1960s and early 1970s was the apparent 'interference' of the third party in the computer section or bureau who constrained the operation, often unintentionally, of the designer or surveyor. This independence however does involve the professional with a keyboard and for the sake of economy of his own time he would be well advised to learn the simple technique of touch typing before acquiring the habit of using two fingers to operate the keyboard.

The personal microcomputer

Now that powerful machines cost less than one man-month in the office it is possible for all interested professionals to have a computer of their own. Indeed some firms are working to this end and they foresee their personnel actually developing the programs, thus giving them control of the software as well as the hardware, for their own particular requirements. This book attempts to give these people a very basic introduction to the use of these smaller machines and how they can be programmed.

There are dangers however. Computers can often become like a drug and it is difficult to wean the enthusiast on to the more mundane tasks of the office. Consequently time is spent writing programs which are not of a personal nature and which a professional programmer could write for the benefit of all the office, much more efficiently, in a much shorter space of time and for less money. In other cases programs may be written for problems which are really not suitable for computerisation but merely take the fancy of the enthusiast. This is sometimes a difficult judgement to make, as a program which is not viable on its own may well be very useful when incorporated into a system containing a number of programs. Some of the 'starter pack' programs in section III come into this category.

By and large there is a great deal of satisfaction in writing and developing your own program. Home produced software may not be as polished as the professional product but at least the operator knows what is in the program, how it works and what the limitations are. This adds to the sense of control afforded to the operator and gives him confidence in its operation. Many of these small machines are more powerful than the main frames found in some offices in the late 1960s. Their potential is often underrated even by those who should know better, e.g. some of the main frame specialists. The trend appears to be away from main frames to microcomputers and it would be a brave man who dismissed the use of micros in the building professional's office.

The future

Where these advancements will end it is very difficult to say. The silicon chip has transformed many of our manufacturing industries. It is doubtful whether an industrial Western society could operate without the support of the microprocessor. It is used in accounting systems, power supplies, food production, transport, medicine and armaments, and is now entering the home. Upon its shoulders a number of 'think tank' commentators have placed the survival of the human race. As we run short of the non-renewable resources of the world the distribution, allocation and use of what is left plus the search for alternatives will require an advanced form of technology to support the political strategies adopted.

Whether 'the silicon chip' will have the same impact on the design professions as it has with manufacturing industry it is difficult to say. In terms of office processes there is no doubt that it will. Loose paper may well become a rarity as communication links via computer screens are developed. Information files will no doubt begin to be replaced by disk files or another recording medium. Even reference volumes, data bases and other sources of information will be transmitted by screen and only conveyed to paper when absolutely essential.

Already the UK Royal Institution of Chartered Surveyors (RICS) Building Cost Information Service has announced that it will be providing a centralised computer based cost information service in 1983 that will be accessible through a telephone line linked to a firm's microcomputer. In another field, technical literature currently circulated on microfiche can be selected through a computer program and displayed on a screen. The UK National Building Specification is available on two small microcomputer disks and will soon be joined by other standard libraries.

In this book we are primarily concerned with building appraisal but this inevitably brings into account the vast store of data required to support the techniques. Arising from the improved, easily accessible, data banks will be a large range of analytical methods for investigating and using the information. These techniques may bear little relation to current modes of thought and this transitional period may be traumatic as one paradigm is substituted for another.

The gap needs to be bridged gently if disillusionment is not to set in once again. For this reason the programs in this book are essentially related to conventional appraisal. The aim is to seek refinement and improvement in current knowledge without a major leap which would be foreign to most practitioners.

In time this will change, and it is the belief of the authors that simulation of human and building performance will develop from a mere research tool, as it now largely is, and will become an aspect of conventional practice.

New models

These new models will be much closer to the reality which they try to predict. To take an example, at the present time an estimating model adopted by the client's cost adviser attempts to forecast the result of another model used by the contractor's estimator. For various reasons related to the difficulty of handling variable data by manual methods neither model has much bearing on reality. We therefore have a complex process of trying to match two systems with each other, neither of which has a close relationship with the manner in which costs are incurred on site.

The machine however will have little difficulty in coping with the large number of repetitive calculations and variability of performance to give a better insight into what can be expected to happen. By simulating from distributions of known performance of site operations, a *range* of results can be derived demonstrating what can be expected to happen in practice. Even strikes, weather, political upheaval etc. can be included as random events.

It is the fact that a *range* has been given, thus recognising the uncertain nature of human performance, which alters the concept of the model. Previously we *had* to use simplistic deterministic data when we did our sums because we didn't have time to do anything else. With the microcomputer it may be possible for us to generate our own distributions and simulations to gain a better picture of the likely range of performance. From an investigation of the range of results we can talk about the *probability* of a client having to pay a certain amount for his building.

What has been said regarding cost will also apply to evaluations of acoustic properties and energy usage. Research has suggested that human behaviour has a greater impact on heat loss than the

U-value of individual components. Households with a dog, for example, lost more energy through air changes than another similar household but without a dog, owing to the greater number of times the back door was opened to let the pet out! It is possible to draw up distributions of human behaviour to account for such factors allowing a more reliable model of heat loss to be developed than was possible with a simple formula.

The data retrieval problem of getting the information from the site face to the machine has still to be overcome. At the moment the process is too time consuming but this will no doubt be solved as machines begin to be used for this purpose as well.

There are dangers in monitoring human behaviour too closely and indeed with the whole concept of computer models. It is very easy to stray into a situation where people are deprived of privacy and oppressively monitored in a manner not dissimilar to Orwell's 'big brother' concept. This inevitably leads us into the field of ethics in computer modelling.

Ethics in computer modelling

Christopher Evans, in his book *The Mighty Micro* (Coronet, Hodder & Stoughton, 1980) had the following to say about the professions:

The erosion of the power of the established professions will be a striking feature of the second phase of the computer revolution. . . . The vulnerability of the professions is tied up with their special strength — the fact that they act as exclusive repositories and disseminators of specialist knowledge.

He then goes on to argue that the computer will eventually allow access to this specialist knowledge for all who require it and therefore the artificial barriers which currently exist will fall. Now this is an interesting idea, but it assumes that machines not only can make judgements but also have the ability to correct their own internal programming when they find they are wrong. This is a different concept to current thinking, where by and large we expect the machine to merely provide support for human decision making.

There is no doubt that computers are, and will be, used for diagnostic purposes. The dangers arise when the computer makes a judgement and the user has little or no knowledge as to how that decision was made. In addition, it may not be as easy as it is with an individual exercising a profession to hold a machine reponsible for an incorrect diagnosis resulting in financial or personal loss.

Differences between manual and computer models

It is therefore worth considering the essential points of difference between a manual model and a computer model. In the list that follows most of the issues relate to the fact that when a computer user operates a program he takes on board the programmer's view of his problem which in some cases may be different from his own. In some cases he may also be adopting the moral viewpoint of the system designer expressed in the machine's logic for decision making.

1. In manual models, decisions as to exclusion and inclusion of alternatives are made by the user. In machine models this is often controlled by the programmer or systems analyst. Examples of this can be seen in the reduction of choice of specification in some computer aided design systems to avoid overloading the memory store.

2. In manual models the user is responsible for the logic, procedure and choices he follows to solve his problem. In a computer program these are often remote from the user and the criteria for choice are sometimes unknown to anyone except the programmer. Computer programs tend to assume a 'right way' of doing things whereas human experience tells us that there are many ways of tackling a problem, the validity of each way depending on the circumstances. There is a danger here, expressed by many writers, that this assumption alienates the technology from the creative human spirit and can therefore become oppressive.

3. The uniqueness of the personal judgement made by the human decision maker avoids large scale error and oppressive impositions on the public. Changes in social need and corrections to assumptions are easily incorporated and the public impact of any error severely limited. Computer programs by their very nature are meant to be used repetitively. Any mistake, bias, or false view held within the program as to how a choice should be made is likely to be imposed on a much larger scale and may indeed be less recognisable.

4. There is also an in-built inertia to change programs when operational because of the time consuming nature of the task. This inertia not only relates to correction of mistakes but also reduces the inclination to monitor the changing needs of society and the program's response to these needs. By the same token there is a tendency when writing programs to constrain the complexity of the model to avoid the difficulties in writing or using the program. Where this occurs it very often means that the human brain can probably do the job better. Expedience can often

become the order of the day to the detriment of the individual.

5. In manual systems it is easier to retain complete or restricted confidentiality than it is with multiple-user computer systems, unless a system of passwords is devised. However passwords tend to add additional overheads to the system, a real problem in a microcomputer, and also add to the feeling of remoteness from the machine.

6. In manual systems human judgement on *qualitative* aspects can be brought alongside those factors which can easily be *quantified*. This is much more difficult to achieve on a computer without a considerable degree of human interaction, which to some extent goes against the *raison d'être* for the machine. Roger Gill of Bristol University, Bristol, England, when discussing the limitations of the systems approach to design, said

> An attempt to model what cannot usefully be modelled could be harmful. . . . There is always present the temptation to concentrate on those systems which can be readily quantified and to shy away from those subsystems which no science can as yet fathom: inevitably a model once made will give weight in a particular direction.

Accepting the view that the computer is here to stay, we then have to ask what can be done to avoid some of these problems arising.

Draft rules for program writers

It would be naive to suggest that a comprehensive set of rules could be written which would overcome the difficulties outlined above. In any case, a set of rules is likely to fall into the same trap as the man writing the program. They will almost certainly be constraining, sometimes not in a constructive way, and possibly become oppressive and manipulative. However the following rules are put forward with an awareness of their inadequacy and lack of preciseness. The authors will be pleased to hear of further suggestions which may provide a better foundation for modelling, avoiding the creation of oppressive tools in the long term future.

1. Confidential information will not be entered into the machine without express permission of the prime source of that information.

2. All project information will be made confidential and secure within the machine, unless express permission to the contrary is given by the prime source of that information.

3. The process of decision making within the program is to be made explicit, through the program handbook, a teaching program on the machine, or some other acceptable means.

4. The criteria for decision making and the assumptions included in a program are to be made explicit.
5. The limitations of choice are to be made clear and, wherever possible, the opportunity to extend that choice (e.g. outside the standard data base) is to be provided by direct manual input.
6. The interface between man and machine is to be preserved at each level of decision making, unless the user has expressly waived this opportunity to influence the decision making process or logic.
7. If a choice has been made by machine logic alone, and this has materially affected the result, then the fact should be recorded on the output.
8. All programs will be reviewed for logic, content, and data on at least an annual basis.
9. Where the computer output relates to a forecast involving an unknown factor then wherever possible a probability range of the forecast should be given or the assumptions stated. This is a protection against the false sense of confidence often given by a single machine output.

Fortunately most of the models currently in use for building appraisal do not transgress these rules. What is really being suggested is that wherever possible the professionals provide good reasons for their acceptance of computer output and make their reasons explicit. In this sense Christopher Evans's view of the decline of the profession may come true as more of that knowledge becomes available to the public. However it is likely that clients will still want to hold someone responsible for any problems that might occur, and herein lies the strength of the professions. They will take the risk for errors arising from whatever the source on which they based their decision, including the computer.

Concluding remarks

The previous discussion on ethics may seem rather academic to many practising surveyors. It has been included as a warning shot which should be heeded and taken into account as our computers become more powerful and our programming becomes more complex. At the current state of the art, at least in building and design, the matter is not critical. In future it may be a different story.

It has been impossible in this book to do justice to the rapid development and changes brought about by the advent of the micro-computer. It is certain that this invention will change the pattern

of all our lives and it is to be ignored at our peril. We shall no doubt
see shifts in importance between one profession and another in the
coming years but it is hoped that all will use the machine wisely for
the benefit of all mankind.

CHAPTER 2
Problem-solving: man or machine?

One of the most popular descriptions of the computer that has been offered in recent years is that of 'electronic idiot'. On the face of it the use of this term may well be appropriate. If you asked a machine to recognise your bank manager when he came through the entrance doors of your office you will find it safer to rely on the vision of your secretary for an early warning rather than wait for recognition by the machine! Recognition of complex moving shapes is not the strong point of a computer. It may well be able to discern after some time the pattern of his face and match it to an image held in its memory. However, a change of hair style, unusual tilt of the head and other variations will almost certainly cause problems and he will be in front of your desk before you reach the fire escape! On the other hand if you asked the machine to add up one hundred four-figure numbers, calculate the arithmetic mean and the standard deviation it will give a correct answer in less than a second. A human genius in mental arithmetic would find it rather difficult to match such a display of expertise.

To take another example, a man was once reputed to have asked a computer which of two malfunctioning watches was the better. The first watch was always five minutes slow and the second had come to a permanent stop. The computer replied that the second watch was obviously preferable because at least it was exactly right twice a day whereas the first was never right at all! This rather trivial example demonstrates that without some complex programming it would be very difficult to give a machine the *context* which would enable it to answer questions of this nature in a satisfactory way. Judgements of this kind involve compromises which by the very design and operation of the computer are difficult to make. Ask the machine to give you the day of the week for 4 July in seventeen years time or a comparison of Greenwich Mean Time with the time in Sydney, Australia, and it will have no problem, providing it has been given the right instructions to follow. To find the answer it can access a set of unambiguous rules and reach a conclusion. Even if rules were not available it could hold the dates for the next twenty years together with the day of the week that each date falls upon, in its memory store. Without fail it could find the relevant date and give the relevant day of the week. Unlike a human it has no problem in recalling vast quantities of information that it has been told to remember.

Machine intelligence

So, is the machine really an idiot? The answer depends on the task it is undertaking. In essence the computer demonstrates aspects of intelligence but is unreasoning. The description 'electronic idiot' has been coined as a defence mechanism against the threat that many people feel when faced with the programming or use of the machine for the first time. The older machines with their flashing lights, special environment and new jargon certainly encouraged a feeling of awe. Fortunately, microcomputers of the new generation are more like the everyday objects known in the office. Even some of their names have been chosen to make them appear friendly and more subservient, such as PET and Apple. The right approach to the machine is to consider it as a tool which can assist in some of those areas where humans are weakest and then to underpin those judgements which the professional consultant is called upon to make.

If we are to understand where the maximum assistance can be given to the building professions then it is useful to briefly consider the differences between the way in which humans and machines deal with problems. This should identify which applications are suitable for machine operation and also where our initial effort and development should be expended. Some of the limitations will be found within the design of the machine and therefore a look at its internal workings is required. These limitations are a function of computers in general but in addition the smaller memory size and slower speed of microcomputers introduces further problems which will be dealt with later.

It is important to reiterate, however, that the difference between the machines developed before microelectronics and those developed since is largely one of size and degree. The transistors, capacitors and resistors have not fundamentally altered in concept, but instead of being made separately and then wired together to form the circuit they can now be produced more or less in one operation on a chip of silicon. The complexity remains, but the components can now be mass produced using computer aided design and microelectronics with a far greater degree of reliability.

The components of a microcomputer system

Viewed externally the building professional's microcomputer system will usually have the following clearly identifiable parts (fig. 2.1):

1. A keyboard and/or digitiser
2. A screen
3. A tape and/or disk unit
4. A printer and/or plotter
5. A control unit for processing information.

Of these components the first four are items required for the input or output of information, while item 5, the control unit, is where most of the computing operation takes place. In many cases, particularly with small micros, the screen, keyboard and box unit are contained within one case. The term 'micro' is used here and throughout the book as a short version of 'microcomputer'.

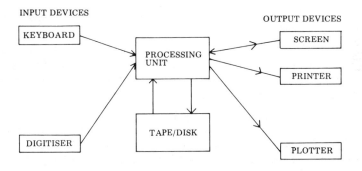

Fig. 2.1 The major components of a microcomputer system.

We shall deal with the input and output devices, or 'peripherals' as they are generally known, in a later chapter. It is sufficient to say at the present time that the input devices correspond roughly to the human senses and the output devices to the way in which we communicate information. It is the processing unit, usually known as the 'central processing unit', or CPU for short, which most closely resembles the brain and which we shall concentrate upon here. It consists of four main parts:

1. *Control unit*. This is the part which controls the sequence of operations according to a predetermined series of instructions i.e. the program.
2. *Store or memory*. This consists of a large number of 'locations', pigeon-holes if you like, where information and instructions are stored. Each of the locations has a unique 'address' so that it is possible to find it quickly and use the information held within it.

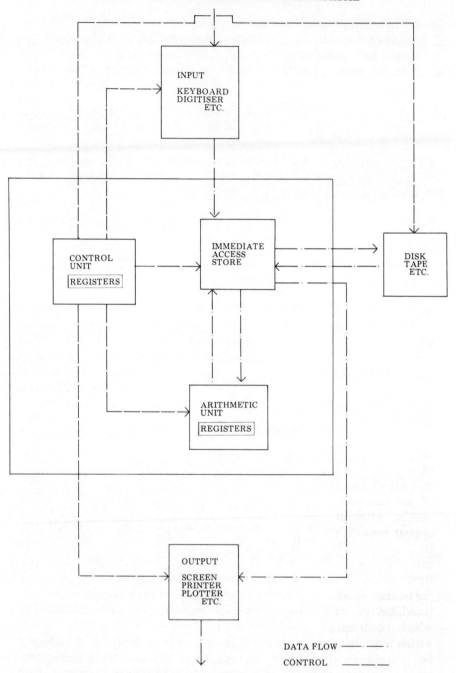

Fig. 2.2 Relationship between CPU components and control and data flow systems.

3. *Arithmetic unit*. This can be described as the electronic calculator of the computer. It works by breaking down the calculations into the simplest arithmetical processes, e.g. multiplication can be done by repeated addition of the number to be multiplied. It undertakes these simple tasks so rapidly that alternative methods are not beneficial. In addition the unit can make comparisons between two alternatives, e.g. one number being larger than another, thus providing the basis for a series of logical instructions based on YES/NO or TRUE/FALSE answers.

4. *Registers*. These are small stores held within the control unit and arithmetic unit that hold the data as it is processed in a calculation and give it up as instructed.

Fig. 2.2 shows the relationship between these components and the control and data flow systems.

Some concepts of computing in general

It is stating the obvious to say the microprocessor is an electrical device. Like any electrical device it works using a series of switches which send pulses along lines of communication between the various components. At any point in time it assumes that there can be only one of two pieces of information being communicated — either there is a 'pulse' present or 'no-pulse'. With this simple piece of prime information the machine can hold numbers (in binary form) and by utilising these numbers as a code can hold symbols such as the letters of the alphabet as well as instructions for the computer itself, and so forth and undertake any number of operations in a variety of sequences. Each 'pulse' will last about a millionth of a second and will affect the information that is held within the registers and stores of the machine. It is this exceptionally fast speed of operation that makes, what in human terms would be a cumbersome operation, appear reasonably efficient. If you can imagine a gate at the entrance to each register or store then the control unit will decide whether the gate is open or shut. It therefore controls what information or data flows along a particular path into a particular register or into a particular location in the memory store in accordance with a defined set of instructions known as the program. The information which controls the machine consists of a series of electrical pulses which can be present or not present. In numerical terms it can be described as having two possible digits, either 'pulse' represented by a 1, or 'no-pulse' represented by 0 (zero). This is known as a binary system. Although the information we may feed into or receive from the machine through one of the input or output devices may appear as normal language or numbers, the machine will have dealt with that information in this 'binary' form. It will have translated the keyboard instructions into zeros (no-pulse) and ones (pulse), manipulated them in accordance with the program and then retranslated the binary information back into alphabetic or numeric form.

Binary arithmetic

An illustration of the form that a number would take in the machine may help to clarify matters. Our number system is based on the unit of ten and is known as the decimal system. (Some practitioners will remember with fond affection the duodecimal system for measurement based on the number 12.) Under the decimal system if I write 6327 it conveys that this number is equivalent to:

$$(6 \times 1000) + (3 \times 100) + (2 \times 10) + (7 \times 1)$$
or
$$(6 \times 10^3) + (3 \times 10^2) + (2 \times 10^1) + (7 \times 10^0)$$

Under the binary system, however, if I write 1110 it conveys that this number is equivalent to:

$$(1 \times 2^3) + (1 \times 2^2) + (1 \times 2^1) + (0 \times 2^0)$$
or
$$(1 \times 8) + (1 \times 4) + (1 \times 2) + (0 \times 1)$$
or
14 in decimal terms

The ones and zeros can be represented electronically in the machine as 'pulse' or 'no-pulse' and the number stored. Binary numbers can be added, subtracted and so on and while humans may find it more difficult to handle numbers in this form, the machine will tirelessly send pulses controlling the electronic 'gates', to manipulate and store the information.

Comparisons between humans and computers

This pulse/no-pulse (or open/shut, true/false, one/zero) form of communication is exceptionally useful and together with the speed of the electrical pulses has provided us with the computer power that we have at our disposal today. However it has also led to some of the limitations of the machine compared with the human brain. It means of course that the whole of the problem must be defined in a YES/NO manner. There are no 'shades of grey' and while this may be a strength in some instances, such as the accuracy required in the calculation of numbers, it can cause a headache when making value judgements where there is a degree of subjectivity. Imagine having a discussion with a client on his assessment of a new building when all he could answer to your questions was yes or no or 'true or false'; or try and define all the factors you would need to consider in order to forecast the cost of a building in two years time by posing a series of questions requiring only YES/NO answers. Although it is possible to overcome the difficulties to some degree by numerical analysis where quantitative data is available, it would be very difficult to describe feelings of comfort or mood of the construction market without resort to some artificial scale of values. The sheer number of questions that would need to be asked in order to arrive at a satisfactory answer would be enormous. These issues would no doubt be decided in a better way by humans external to the machine who could then input the results of their deliberations into the machine for it to operate upon. By using a YES/NO, TRUE/FALSE, process for the internal logic of the machine it means that the problem must be structured without any form of ambiguity. Everything must be stated clearly and precisely in simple terms. A person may have difficulty in describing an elephant but would not find a problem in recognising one! Yet the machine would not recognise an elephant, or a building, or a flow of heat, or a cost prediction problem, without it being accurately and uniquely described to it. The old joke of, 'Can you tell the difference between a pillar box and a camel?', to which most people reply, 'No', and the questioner replies with, 'Well I won't ask you to post this letter then!', has some relevance to the computer. It would need a very precise description before it actually could tell the difference. If it were considering a building then the dimensions, co-ordinates, orientation, thicknesses, texture and so on would need to be included, together with the fact that it *was* a building, before a proper evaluation could take place. This description of the building would of course need to be held in binary form.

To assist in the clarification of problems to suit the binary system it has been found that Boolean algebra can make a major contribution. This method (named after the mathematician George Boole, 1815—64) uses algebraic notation to express logical relationships. It would not be appropriate to explain the concepts in a book of this nature. It is sufficient to say that the relevance of Boolean algebra to the logic of computers lies in a simplification of the system in which the values of the variables are restricted to two possible 'truth values' of a statement, i.e. 'true' or 'false'. These values may be represented by the digits 0 and 1, thus enabling the logic of Boolean algebra to be applied to the binary logic of computers. Where there is ambiguity it will inevitably lead to error. As a word of encouragement, however, all members of the building design team are well versed in the preparation of legal documents, specifications, bills etc., and this is good training for the rigorous structuring and definition of problems required by the machine.

The problem context

Related to ambiguity is the question of cross-referencing. Whereas a person can immediately place a space or component into its proper context within the building, identify its repercussions on other spaces or components, and assign attributes of cost, colour, thermal efficiency and so forth, the machine needs to be told. It will not account for any aspect which has been omitted nor will it make realistic assumptions until some instruction has been included to tell it to make that assumption. The properties of any component will need to be spelt out in detail as the computer could not associate certain attributes with certain materials unless it had been told to do so. Thus associations we take for granted need to be made explicit in the machine. At this point in our discussion it is worth noting that a common way of undertaking this is to store as a group both the component and its attributes in the memory. When the component is taken from a store the properties come with it and those that are relevant to a particular evaluation are abstracted by the program. This is rather like the manner in which the brain stores patterns of the world it discovers around it and uses these patterns to make sense of what it sees. Providing the patterns can be defined in a precise manner they can be fed into the machine and the computer can cope. Unfortunately it is not always possible for a person exercising a profession to be able to know and define all the factors and patterns that he takes into account when making a 'professional judgement'. The criteria he chooses may well have been gained from experience, without formal learning, and this provides an 'intuitive feel' for the judgement in hand. It may be extremely difficult to spell out these factors in an unambiguous manner. Indeed if it were simple to do this then the number of contractual disputes would fall rapidly!

At the present time a good deal of research is being invested in what is known as 'expert systems'. Interest to date has been largely from the medical profession where the diagnosis of illness by experts is fed into the machine together with the symptoms and other factors they consider relevant. The machine 'learns' from the consultant what points of reference should be employed, what combination of factors, what criteria should be adopted, in order that a correct diagnosis can be made. Other less able practitioners are then able to access this information for the benefit of their patients. There may be applications here for the building professions but it is certainly outside the scope of this book and the machines we are considering at the present time.

Limitations of memory

Another limitation inherent in the machine is the memory capacity. Unlike computers the human brain receives a complex series of multiple inputs from the body senses. The combination of these senses tell us what the world is like within the range of what they can detect. The brain has a finite capacity and therefore to cope with all this information it uses screening, unlearning, pattern making and other techniques to sieve and retain what it wants to know, in order to operate to its satisfaction. It will even classify information into short and long term memory requirements so that transient information can be held only as long as it is required. However the short cuts used in these techniques, and the lack of regular reinforcement for the information held, will occasionally and sometimes frequently result in error. The computer on the other hand will accept all the information it is asked to remember until its capacity is exhausted. The programmer will have to consciously decide to remove, draw associations between, analyse, structure and classify the information held. However the memory capacity, particularly in microcomputers, is very much smaller than the human brain. Consequently all the associations a human can bring to bear upon a problem cannot at present be fully represented in the machine. The memory that is found within the central processing unit can however be supplemented by 'backing store' usually consisting of tape or disk. This additional memory is very similar to the kind of remote but accessible knowledge we find and use in books. We cannot remember everything so we access complex factual information only when we need it by looking it up in a reference volume.

Similarly the machine draws into its main memory a program or some data from the tape or disk only when it requires it. This means that the main memory store can be used more effectively, longer programs can be run and storage space is not wasted. The time penalty in using a disk backing store is nothing like the time spent in using a book as transfer is extremely rapid — usually within a second or two. Tapes can however be quite time consuming as information has to be stored sequentially and the tape searched at a fairly slow speed to find the relevant data. The machine does however have one very big advantage over the human brain — it is extremely accurate. Any error that is likely to occur will almost certainly emanate from a human mistake. As the machine itself can be thoroughly tested with diagnostic equipment it is unlikely that major faults in the components will arise in manufacture. The problems that do occur will be in the links between the components, the moving mechanical parts and the instructions used to access the memory contained within the program.

The program

The program is merely a set of instructions written to enable the machine to perform a particular task. In order that the machine will understand the program, the instructions will have to be given in binary form or machine code. However because this is difficult for humans to write, programs are usually written in one of a number of standard computer languages which approximate more closely to English. A 'compiler' or 'interpreter' is used to convert this language into binary form called 'machine code'. What the program will do is to follow through a set procedure executing a series of tasks to solve a problem. The machine will slavishly hold to that procedure and without fatigue continue to execute the instructions until told to stop. The psychological problems associated with repetitive human work no longer apply and therefore those manufacturing and commercial applications which are of a repetitive nature are liable to be subject to machine takeover. Some professional tasks are of this nature but quite a number of professional activities involve creative acts or judgements which cannot always be defined in terms of a linear process of decision making. Adapting these latter activities to the machine creates some difficulty. The nature of a program is such that a set pattern of approaching the problem is assumed. The designer and his colleagues may thus be forced into working in a particular way which may be to the detriment of the decision that he is trying to make. This is particularly true of architecture where the designer may be trying to balance in his mind such things as the organisation of spaces in the building, the external appearance, legal and technological constraints, and cost. The multiplicity of options and criteria for making what sometimes appears like a simple choice are enormous. To develop a program which asked the architect to examine each aspect in turn in a set way would almost certainly be detrimental to arriving at a successful solution. However it is possible to write the program in such a way that the approach to the problem can be varied so that the emphasis required for a particular project can be reflected in the procedure adopted in the machine. The evaluation routines can be held as sub-routines (self-contained parts of a program) which can be accessed in whatever order the user requires. There will be more discussion on the procedures and logic of programming in chapter 5 and section II of this book.

Input and output of information

The last major difference between computers and humans is the manner in which they receive and output information. The human brain is constantly receiving information through what we call our senses. These include our sight, hearing, touch, taste and smell. These senses acting together through the brain not only form a picture of the world but also help in identifying and structuring the problems needing to be solved. The computer on the other hand is usually fed information a piece at a time (albeit often at very high speed) by an electronic device and the problem has to be externally identified, structured and 'solved' before being held within the computer program. The input to the machine must therefore he highly organised and structured to suit the program. The devices for input are extensive and include keyboard, magnetic tapes, voice recognition, optical devices for character recognition, light pens working in conjunction with a cathode ray tube, and digitisers (a device which can convert a physical quantity, e.g. the measured length of a wall, into x, y co-ordinates for calculation purposes). The computer can also be attached to analogue devices for measuring electrical energy, heat, light, sound etc. These input devices operating with the central processing unit do not compare at the present time with the organisational power of the human brain within the range of its senses. However, as some of the devices will measure more accurately than the human brain and limbs will allow, and will also extend beyond the capability of the human senses, then the *range* of inputs can be said to be more extensive.

Output devices, on the other hand, are another matter. The nature of human communications through the voice and through limbs which can operate tools such as pencil, pen, typewriter, switches, knobs etc. is exceptionally slow compared with the computer. Even the fastest typist could not keep pace with the slowest of the currently available computer printers. Nor could he or she hope to cope with the degree of accuracy achieved by the machine. The normal methods of outputting information from the computer are by means of the VDU (visual display unit — rather like a tele- vision screen), a printer or plotter. When the output information does not require to be immediately understood by the user it can be output to a magnetic tape or disk where it is stored for future use. Disk operation is particularly fast with a writing and reading speed of several thousand characters per second. Once again, how- ever, the nature of the machine is such that the output must be highly structured as part of the computer program and therefore the flexibility that we have as human beings to construct language, adjust language, and tabulate etc. to suit the requirements of the project in hand, is not easy to achieve with the machine. The program is once again the source of the type and form of output coming to and controlling the output devices.

Full potential of the machine
The above discussion has skated across some of the major differences between the capabilities of machine and humans. Professor Herbert Simon of the University of Chicago has stated in *The New Science of Management Decision* (Prentice-Hall, 1977), however, that

There is nothing about a computer that limits its symbol-manipulating capacities to numerical symbols; computers are quite as capable of manipulating words as numbers. In principle, the potentialities of a computer for flexible and adaptable cognitive response to a task environment are no narrower and no wider than the potentialities of a human. By 'in principle' I mean that the computer hardware contains these potentialities, although at present we know only imperfectly how to evoke them, and we do not yet know if they are equivalent to the human capacities in speed or memory size.

If these assertions are correct then it may be that the apparent limitations we see around us today will be absent in the long term future. A true artificial intelligence (said by one researcher to have arrived when a 'computer sociologist' is required to analyse the machine performance rather than a systems analyst/programmer) will create enormous ethical and social problems. Fortunately it is a long time away and well outside the objectives of this book.

A summary of the preceding points is included in table 2.1. The object of this discussion has been to provide a context for the more detailed consideration of the application of computers to professional problems in building. Without some knowledge of the similarities and differences between the machine and ourselves it is difficult to grasp the potential of the microcomputer in our everyday work.

Comparison between main frame and microcomputers

We have covered so far the general points of difference between man and computers. However with microcomputers there are further limitations which are not so much a difference of kind but of degree when compared with a large central computer. These can be considered under two headings; memory capacity and speed of operation.

Memory capacity

INFORMATION HELD IN MEMORY
Micros generally have less main memory capacity than a main frame. At this point it is worth-while considering what actually needs to be

Table 2.1 Summary of comparison between man and computer (with reference to G. Broadbent, *Design in Architecture*, Wiley, 1973)

	Machine	*Man*
Speed of operation	Extremely fast and superior to man in most respects	Slower at physical and simple mental work
Stamina	Can sustain operation *ad infinitum* and limited only by reliability of mechanical and electrical devices	Unable to sustain long periods of work without rest
Accuracy and consistency	Extremely accurate, consistent and reliable. Poor at error correction	Unreliable and inconsistent particularly when dealing with repetitive work. Good at error correction
Senses	Consistent sensitivity. Finds difficulty in sensing from a variety of sources simultaneously. Wider range of sensitivity dependent upon sensing device	Ability to combine sensory powers. Senses can be affected by environment, drugs etc. and dulled by too much exposure in some cases
Memory	Generally smaller (in micros, very much smaller) but very accurate storage and recall	Large memory but subject to memory loss and error in recall
Overload	Sudden breakdown	Gradual degradation
Cognitive ('knowing') processes	Follows instructions with complete accuracy, performing logical and/or arithmetical operations	May follow instructions precisely or in a haphazard fashion. May misunderstand them
	Logic largely a matter of determining whether a statement is true or false. Therefore mainly a process of comparison with given information	Processes may be interrupted by 'creative leaps' short circuiting a tedious procedure
		Logic may be suspect
	Man made criteria required for judging whether true or false	Perceiving, remembering, imagining, conceiving, judging, reasoning affected by feelings, emotions, motivation and so on
	Good at arithmetic operations but can only draw and compare simple analogies	Good at comparing and judging unlike things, at inductive logic and at drawing analogies

Table 2.1 contd.

	Machine	*Man*
		metaphors directly relevant to the problem in hand
		Good at making complex decisions
Input	Machine input exceptionally fast but human input slow	Slow at character reading but fast at sensory recognition
Output	Very fast, neat and tidy. Output to backing store exceptionally fast. Lack of flexibility in manipulation	Slow output at all times either by speech or hand. Very flexible in manipulation

held in the memory of the machine in order to provide a useful tool for the building professional's operations. The information required includes:

(i) *System data or operating system.* This is the program which brings the computer into operation when switched on and continues to control the various hardware (machinery) and software (programs) activities. It is this program that will determine the ease of operation by the user, the flexibility within the system and to some extent the speed of operation (particularly with regard to input/output devices). In microcomputers this information is usually found in the read-only memory of the machine (see page 36).

(ii) *Program instructions.* These are a sequence of steps or operations which describe the route through the problem-solving process which is understandable to the machine. These operations will include such things as mathematical calculations, logical 'jumps' to various parts of the program based on true/false assumptions, and the format required for the printed output of information. On microcomputers this type of information is usually fed into the machine through the keyboard and then recorded on to either a tape cassette or disk. Whenever it is required again it is simply read in from one of these devices into the internal memory of the machine.

(iii) *Changing but recoverable data.* This might be information fed in by an operator via a keyboard as part of a program. If the computer loses the data it can always be fed in again.

(iv) *Changing and irrecoverable data.* This is information over which the operator has very little control and which would be difficult or impossible to recover if lost. If a program has a 'looping' routine (whereby a series of operations is repeated a number of times) then the machine will need to count the number of loops or iterations. The count will need to be recorded in the machine but it may be difficult to retrace the steps to the relevant count number should something go wrong.

(v) *Intermediate data.* This is data generated within the machine itself and which may be used again later in the program. It will include the results of calculations and user choices which will affect later aspects of the program operation.

In addition to the above, standard information may also be held including such things as the number of days in a month, the time recorded by an internal clock and standard sub-routines to calculate mathematical functions. If the machine is to be dedicated to one particular task then the memory will have to hold the fixed program for that task. This application is however more appropriate to manufacturing work rather than the work of the design or building consultant.

TYPES OF MEMORY

It is now necessary to introduce some further jargon to consider the types of memory available:

(i) *Serial and random-access memories.* With a serial memory (found on simple backing store) the data is stored in sequence in a fixed order. Magnetic tape is an example of this type and when placing new information on a cassette it must follow on from previously held data or overwrite it. To find an item of information it may be necessary to search the whole of the tape until the item is discovered. This is naturally very time consuming — imagine starting at the beginning of a telephone directory and looking at every name in sequence before you discovered the name you required! The advantage of serial memory is that it usually offers large amounts of storage space at very little cost but it has the penalty of the time required to access a piece of information.

With random-access memories (such as that found in main memory), on the other hand, the storage elements are each given a unique address. It is therefore possible to access any piece of stored information directly at any time, with very little delay at all, merely by calling that address.

(ii) *Volatile and non-volatile memories.* A volatile memory keeps the information held within the machine only while there is a constant supply of power. If the power is interrupted, say by a plug being accidentally removed from its socket, then the contents of the memory are completely lost. In contrast a non-volatile memory is permanent and independent of any power supply. Consequently any information which is to be used repetitively should always be held in non-volatile memory and copied to a volatile memory when it is more appropriate for the operation in hand. An example of volatile memory is the internal memory of the machine whereas an example of a non-volatile memory is the disk or tape storage external to the central processing unit.

(iii) *Read-only memories (ROMs)*. Data is placed into memory by 'writing' it and is retrieved by 'reading' it in computer parlance. A read-only memory therefore is one whose stored information can be retrieved but cannot be easily altered. The purpose of the ROM is to hold permanent data or information that will not change and which must be retained in the memory even when the power is switched off. In a microcomputer the ROM's main function is to hold the program of instructions which tells the microprocessor what to do. There are a number of different types of ROMs such as EPROMs (erasable programmable read-only memories) and EAROMs (electrically alterable read-only memories) but these tend to be used for specialist work dedicated to a particular function. In general terms, ROMs are non-volatile memories used for the storage of permanent instructions such as the operating system data and standard functions outlined previously.

(iv) *Random-access memories (RAMs)*. This type of memory is used where data needs to be written to or read from memory for the purposes of the computer program. The term 'random-access memory' is a confusing one since ROMs are also random access. It is perhaps better to think of RAM as 'read and modify' to give a clearer definition. Because data can be read to store, changed and retrieved at will, this type of memory is the most versatile. However most RAMs are volatile so that, as noted previously, if you lose or interrupt your power supply the data is lost. They are used for such operations as holding the program, intermediate storage and storage of the results.

MEMORY STORAGE SPACE

Now it can be seen that the storage space required in the machine is not merely the space needed to just hold the program. There must also be spare capacity to cope with the original data, intermediate results of calculations, records of choices made, counting mechanisms and so on. This additional data can well require storage capacity in excess of the program instructions. This is particularly likely to happen when a routine is to be repeated a number of times and the results of each iteration are to be held for output or analysis. Unfortunately the main storage capacity of microcomputers at the present time is fairly small compared with large main frame computers. The most common memory sizes at the present time allow between approximately thirty-two thousand (32K) and sixty-four thousand (64K) characters to be held in the internal main storage. This is sufficient to run a good sized program providing large quantities of information are not generated which need to be stored. It is possible to overcome the problems of limited main storage if backing store in the form of disks is available. The program can be divided into sections and dealt with in linear fashion, i.e. one section being operated upon before the next section is entered into the machine and executed. This would require a predetermined order for the sequence of operating the section 'blocks'. The new block would be overwritten into the part of the memory holding the previous piece of program but would use the same names for the data locations and therefore be able to pick up the results of previous sections. Not all programs can be divided up in this way and in any case the 'overlays', as they are called, do become a rather cumbersome feature in the program and this slows it down and decreases the flexibility of operation due to the 'blocking' procedure. Much larger capacity micros are already on the near horizon and in due course there is no doubt that this particular limitation will be overcome. The problem lies in the amount of information that can be held on the silicon chip and the number of wires connecting the memory to the chip (known as the address and data buses). Experiments are being undertaken with materials other than silicon and increasing the size of the address bus (i.e. electrical channels) and therefore these particular constraints are likely to disappear. Main frame machines do not have these limitations because they use a larger address bus and a number of different chips to handle the memory storage, each of which has a specific function. This makes the handling of information more efficient and more memory can be added on. Micros attempt to get all the memory for their central processing unit on one chip. The total capacity is therefore limited by the number of wires on the address bus, usually 16, giving an absolute maximum of 2^{16} or 64 kbytes of memory store.

The memory capacity of main frame machines is usually several times larger.

Speed of operation

When compared with a main frame a microcomputer is slow. It probably seems to many lay people that those who are concerned with the design and application of computers are worshipping speed for its own sake. If a machine can perform a million additions in a second, why try and go any faster? In fact why go for high processing speeds when the peripheral devices (e.g. the printers, disks, etc.) have the greatest effect on the speed at which most commercial jobs operate?

ADVANTAGES OF SPEED

These points are valid where the machine is merely taking in a limited amount of information, executing a simple mathematical routine and outputting a result. There are however occasions when speed of processing is important and these include:

(i) *When a large number of iterations are involved.* An example of this type of program is where there are repeated simulations of activity, e.g. pedestrian behaviour in buildings, a production process and cost simulations. These techniques tend to be used and developed by researchers and they are therefore unlikely to become a problem to the normal practitioner at present. However as program packages become available arising from this research then speed will become important in this type of work.

(ii) *When a large amount of data needs to be entered, stored, sorted and analysed.* Such is the case with bills or schedules of quantities where thousands of dimensions are entered into the machine together with a large number of descriptions or codes. The machine is required to group like items together, undertake the calculation of quantities, sort into order, and call up headings and so on before printing out the final document. These processes involve many times more computer operations than the actual number of entries. Processing speed becomes of vital importance if the machine is not to be tied up for many hours with the continual supervision that is required.

(iii) *When the response time is psychologically important.* There are a number of occasions when a delay in the machine replying to a command or request will result in either boredom on the part of the operator or worse still a lack of credibility on behalf of the user and his colleagues. Imagine waiting ten minutes for an answer from the machine in the middle of a site meeting! It may be a good excuse for a coffee break, but if it happens too often the disruption will force the building team to consider whether it is worth while! Many other professional activities are time dependent, having to be worked within a normal eight-hour day for example, and therefore the flexibility to extend beyond normal working hours on an irregular but frequent basis does not always exist.

(iv) *When unskilled programmers are being used.* A fast efficient machine will very often make up for inefficiencies found within the program. A highly skilled programmer will develop techniques and routines that will save on memory size and speed of operation. If it is not possible for the firm to employ this type of personnel then a faster machine will offer some compensation.

SPEED AND COMPUTER WORDS

At present main frame machines tend to be much faster than microcomputers. One reason for this lies in the way in which they handle what in computer jargon are called 'words'. The machine handles chunks of binary information at a time. A single binary digit is called a 'bit' (i.e. a zero or one) and this is the smallest piece of information dealt with. If we put eight bits together (and this is the number conventionally stored as a group in the microcomputer's memory) then it is called a 'byte'. A 'word' on the other hand is the number of bits a given microcomputer handles at a time. Word lengths usually vary between 4, 8, 16 and 32 bits, so a 16-bit word would contain two bytes (fig. 2.3).

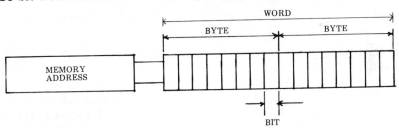

Fig. 2.3 A word in memory and its components.

Most microcomputers currently available at the cheaper end of the market use an 8-bit word, that is they handle a minimum of 1 byte of information at a time. As a general rule, the longer the word, the more 'powerful' the microprocessor. This means that fewer discrete instructions are needed for a particular job, and more memory can be directly addressed. The word length can also affect accuracy. However the accuracy of a machine depends upon the number of bytes used by the operating software to represent a floating-point number. Different machines use one, two or four bytes. For example a single 8-bit word offers 256 (2^8) combinations —

sufficient to give an accuracy of approximately 0.5% in calculations and enabling it to be used for the code representing the letters and numbers input from a typewriter keyboard. This is developed further in chapter 10. In comparison a 16-bit word gives 65 536 (2^{16}) combinations of 0s and 1s, providing much greater accuracy. However an 8-bit micro can often be capable of working as a 16-bit word processor but only at the expense of increasing the number of machine instructions required and consequently decreasing the apparent speed of operation.

Main frame machines commonly use a 32-bit word and therefore the amount of information dealt with in each operation is greater and speed of processing is increased. This increased speed is one reason why the main frame machines can support a large number of terminals whereas microcomputers are commonly limited to just one or a very small number of keyboards. In addition the internal organisation of the main frame machine, which handles all the machine's instructions, is usually more powerful and extensive and this in turn provides a faster turn of speed.

Summary to date

In this chapter we have looked at the factors that tend to limit computer technology at the present time and this may be a good opportunity to summarise the major strong and weak points before looking at the nature of building problems.

The computer's strong points are:

1. Arithmetically reliable
2. Fast arithmetic calculation
3. Fast and accurate recall from main memory
4. Reliable adoption of program logic
5. Fast input and output of information using peripherals
6. Clear and fast formatting of information
7. Not tired or affected in any way by repetitive calculations.

Its weak points are:

1. Fixed program logic tends to lead towards an inflexibility in approach to problems
2. Slow to perceive complex information at one time, e.g. 'recognition'
3. Memory is limited (particularly in microcomputers)
4. Has difficulty in coping with problems with a wide context at speed
5. Cannot compromise very easily
6. Requires the problem to be precisely defined.

These factors set the scene for a discussion on manual problem solving in building and the assistance that the computer can give. While we have been talking mainly about all types of computing in this chapter it should be stressed that the microcomputer will only add further limitations when it comes to the size of memory and the speed of operation. However these two factors will in turn produce more constraints and in particular the range and choice of peripherals able to be added on for use in solving building problems.

The nature of the consultant's problem

It would be impossible in a book of this nature to provide a comprehensive view of problem solving for the client's advisers. Indeed it would be true to say that as yet there is no real consensus view as to 'how' the design problem should be structured and tackled. It is possible, however, to identify the goal of the building professional and to suggest areas of common ground in the way 'certain' problems are tackled.

The object of the design team must be to provide accommodation for human activity in such a way that both society and client will recognise the end result as having 'value'. Now 'value' in this context can be defined as achieving the right balance between 'wants' and 'resources'. As 'wants' and 'resources' vary in different parts of the world and between different people, so will the concept of value. Value is not therefore an absolute. Neither can the term 'value for money' have any real meaning, for as money is a measure of a resource it is intrinsic to the word value in any case. (It is however possible to speak about 'quality for money', provided that a satisfactory measure of quality exists.)

Therefore it follows that as value varies from case to case and project to project then a clear definition cannot be laid down. In fact the briefing process, which proceeds throughout the design period and sometimes beyond, is an attempt to achieve the right balance and provide value. It is a complex compromise and attempts to combine a variety of measurable characteristics (heat loss, daylighting, cost, etc.) with an even wider variety of unquantifiable factors (comfort, spatial environment, texture, etc). Many of the so-called measurable characteristics are not able to be precisely determined because they are functions of human behaviour which cannot always be controlled. Prediction of heat loss and cost for example are notoriously difficult to achieve because they rely so heavily on human behaviour and management. Yet the design team are called upon to choose the right materials and level of insulation at a suitable cost in order that value in terms of energy usage is achieved for the client. Even if this were possible (and we probably can make a reasonable judgement) it would be very difficult to say with any confidence that value has been achieved for society or the world. If we wished to conserve the world's scarce energy resources then our simple mathematical models would prove inadequate to cope, and another set of compromises, this time between client and society, would have to be made.

Professional judgement

The purpose of the previous few paragraphs has been to outline the difficult nature of professional judgement. The problem the building professional faces is what Schumacher, in *A Guide for the Perplexed* (ABACUS; chapter 10), calls a divergent one. Any number of skilled personnel can sit down and logically solve the problem to their own satisfaction and yet arrive at completely different solutions. Their solutions will depend upon such things as their work experience, upbringing, prejudices and the choice and weighting of their points of reference. Now bearing in mind what has been said about the present limitations of computer technology it will be recognised that the machine is not able to cope with such a poorly defined set of circumstances involving subjective value judgements. It requires the problem to be clearly identified and structured and to have a written, defined method of obtaining a solution. Compromises are not possible unless these can be expressed mathematically by, for example, the ranking and weighting of priorities. However ranking and weighting, where it is possible, will vary from project to project and no universal guidelines can exist unless you are prepared at the same time to risk the maintenance of a free society.

It therefore follows that, at least for the present, aesthetics, spatial organisation, texture and colour, comfort levels and so on are not the province of the computer. So which problems are suitable? Inevitably they are the quantifiable sub-problems of the major problem, i.e. those aspects for which descriptive and measured information can be produced. This information can then be used in the overall human controlled 'weighing up' process that will be used to produce the end solution to the main problem. These can be described as 'convergent' problems. In these problems the factors which cannot be strictly controlled, or at least, measured and 'allowed for' are removed and what remains is an isolated system containing a soluble problem. Each part of the problem can be identified, ambiguity and compromise are removed, and a method or technique for providing a solution has been established. This does not mean that the approach is perfect but that the technique is known to produce reasonable results when using manual methods. Problems for which a mathematical equation can be written are ideal for machine appraisal. Having said that, it needs to be fairly heavily qualified. It is terribly easy in a flush of enthusiasm for the microcomputer to use it for applications which are still better undertaken manually. There is still a tendency to use a sledgehammer to crack a nut when first using a machine. However in a planned and developed system small progams could be a useful part of a comprehensive package of sub-routines for use in building evaluation. Nevertheless there are some applications even of a mathematical kind which are not in the authors' view suitable for microcomputer operation.

Problems at present unsuitable for computer application
This is where large amounts of data need to be fed into the machine for a unique project before the mathematical operations begin. Memory space is limited on a small machine and therefore handling large quantities of data is a cumbersome exercise as well as being slow. Leaving this aside, there is also the physical problem of entering and checking the data. An example of this type of problem that has already been mentioned is the production of a bill of quantities. Thousands of measurements need to be entered together with description codes and they all have to be checked for error. By the time the measurer has written out his dimensions, searched for his code (usually undertaking a preliminary sorting of the information at the same time), drafted and checked his work and that of the machine operator it may well be quicker to have used a fast manual system such as 'cut and shuffle' in the first place. Unless he can use that information for project control and analysis it is unlikely to pay, although savings on typing and printing may eventually make it very worth while. It would be better to take the successful tenderer's bill, prepared manually, and enter that into the machine for contract monitoring. The machine can then be used as an information and data store which is accessible for techniques such as cash flow prediction, UK National Economic Development Office (NEDO) price fluctuation calculation and elemental analysis. Until such interesting developments as voice input of dimensions and laser scan measurement are fully evolved, it will remain debatable whether the solving of problems which of necessity have large data inputs for single projects is an economical proposition. It is rather like the problem that used to be faced by weather forecasters. From the enormous amount of data available from satellites, ground stations and weather balloons it was possible to forecast with great accuracy the following day's weather. The only problem was that it took two or three days to collect and analyse the information!

Satellite pictures quickly changed the situation. In a similar way a simplification of the rules of measurement could transform the return on the investment in a computer.

General rules for machine application
From the above discussion we can therefore put forward the following general rules:

1. Machines should not be used for making subjective value judgements unless the criteria for value can be well defined.
2. The relationships between the results of sub-problems involving the ranking and weighting of priorities is probably better left to human decision making.
3. Where a large amount of unique data is required for the solution of a single problem then this is probably better prepared and sorted by an efficient manual method than by machine.
4. Unless the problem is very complex, then for 'one-off' projects where only a single solution is required it is better to use manual methods.
5. Where the problem can be well defined, preferably in a mathematical form, and where the routine needs to be repeated several times for the same or different projects, then the microcomputer is a suitable tool for evaluation.
6. Where the problem involves the holding and processing of a large mass of data (e.g. meteorological data or the attributes of a building), then the memory size and speed of the microcomputer will prove to be a severe limitation.

Having looked at the type of problem that should be considered it is necessary to consider the techniques used to establish solutions to convergent problems.

The nature of building appraisal techniques

Most of the techniques employed by architects, cost consultants and the like and which can be made explicit for computer operation have a number of common aspects.

The vast majority of problems with numerical solutions will follow the pattern of fig. 2.4. Data will be collected and structured to suit the model (i.e. the representation which is used to solve the problem such as a formula or some form of measurement). Once the model is constructed and run (e.g. a heat loss calculation or elemental estimate) the results are analysed to discover as near as possible the 'best' solution. This solution is implemented and the results as measured in the final solution are monitored to provide

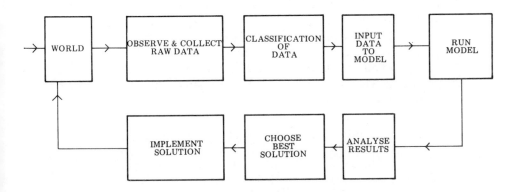

Fig. 2.4 General problem-solving system.

further data for analysis. Invariably there will be a certain amount of looping back through the process to try different options before implementation.

The process is the same whether it is to be followed manually or by the machine. There will however be a major change in the kind of techniques used in the system as machine operation becomes established. Nearly all our current techniques have evolved to suit the limitation of human limbs and brain. This invariably means

   (i)   a reduction in the amount of data to be handled
 (ii) reduction and simplification of calculation processes
(iii) reduction in the number of options to be considered
(iv) simplification of output.

The manual models used have largely formed a skeleton upon which the consultant has hung his professional judgement. The judgement was necessary in a good number of cases because time was not available to construct a better model and manual analysis of raw data was too time consuming and costly.

However with the advent of the machine some of these constraints disappear. Difficult and tedious calculation is no longer a problem, and provided that input of data is not extensive then complexity within the model, as long as it can be well defined, is also no longer a limitation. The fact that the computer can tirelessly record and output data means that, used wisely, more detailed information can be fed to the user. (We have all seen program outputs which bury the operator in paper!) These factors contribute to what is potentially a major advance in the tools available to the building consultant.

Improvements in building appraisal

There are probably four major ways in which the information used in building appraisal can be improved. In each of these the micro-computer can assist, although it should be recognised that the mere fact of introducing the machine will not provide the panacea for all our problems. The techniques themselves need to be developed.

The four ways are:

1. The information can be provided *quicker*.

   In other words the response period, from the time a proposition is made to the time when it is evaluated, is reduced. This has the advantage of cutting down the time involved in the decision making process, possibly allowing other options to be considered or providing the client with the building at an earlier date. The speed of calculation in the machine and the rapid output will obviously assist in this respect.

2. *More* information can be provided.

   If it is assumed that the more information that is provided, the better the decision to be made, then the ability of the machine to tirelessly undertake iterative processes trying large numbers of alternative solutions will obviously be a considerable advantage. Whether the human brain can cope with the rapid increase in data which it has to 'balance' in making a design decision is debateable.

3. *More reliable* information can be provided.

   Here the machine can only assist in one respect. That is, the recall of information and arithmetical calculation of results will be very accurate and far more reliable than the human brain. However this is only one factor in the reliability of information. The trust that can be put in the output information will depend on the techniques used (contained within the computer program) and the input data upon which the results are based. The well known computer maxim 'garbage in, garbage out' is undoubtedly true. In human decision making the consultant can often compensate for bad data by exercising his 'professional judgement'.

4. Provide information *at an earlier stage in the design process*.

   Suppose that the information which is normally provided later in the design process could be brought forward to a point where it was used before major design decisions such as building shape, orientation, height etc. were finalised. It would then be possible to consider these important factors together with the various specification options before the design team had committed itself too far. The right strategy for development of design could be determined and the probability of abortive work and unsuccessful design reduced. It would also allow the client's brief to be firmed up at an earlier stage as the evolution of the client's requirements would be considered in a condensed period of time at the very beginning of the project. With manual techniques this bringing forward of the major decision making processes is difficult to achieve owing to the time consuming manual appraisal. The advantage the machine has is that it can hold the data relating to specification in its files, it can appraise the effect of building geometry very quickly and it can be made to automatically try out a range of options. Although techniques to undertake this work are largely in the development stage, the concept is well founded in a number of computer aided design systems and it is the authors' view that it is in this area that microcomputers will be of major advantage to the design team. The subject is developed further in chapter 4 where office systems are discussed.

CHAPTER 3
What system do I need?

It has to be said that a large number of computer systems have been purchased in the construction industry for the wrong reasons. These reasons include pressures from some enthusiastic staff not always supported by all levels of the firm's employees; an urge to compete with the front runners of the profession; and a desire to be part of 'the future' as if 'the future' had little to do with the past. In the right context each of these reasons may have some validity but on their own they are insufficient.

The critical question to ask is, 'Do I need a computer at all?' In the short term at least, if manual systems are working well, resulting in client, employee, management and profit satisfaction, it may be worth waiting until the technology provides a more obvious benefit than can be seen at the present time. However if there are some indications that improvements can be made then careful consideration of microcomputer technology should be given. Such indications include:

  (i) An inability to respond quickly to the demands and requests of the client or another building-team member.

 (ii) A need for greater reliability of calculations and output.

(iii) Overloads on repetitive work at various times in the month (e.g. price adjustment of contracts for valuations).

(iv) Time consuming recording of information which then cannot be accessed, reproduced and processed quickly.

 (v) A reluctance to undertake too many appraisals of design ideas owing to lack of time.

Some organisations have been able to make the purchase of a micro-computer system pay within a year by using it purely for the calculation of the NEDO price adjustment formula. However such limited use is the exception rather than the rule and a more extensive range of program applications will be required for the majority of users. Having said this it should be noted that the level of commitment to computing required to purchase a micro is nothing like as great as that required to run a main frame or minicomputer system. A £5000 outlay on machine and software soon seems small when tax relief is deducted and a smile is seen on the client's face! It will probably represent less than a quarter of a 'man-year' and with the right

right management this should be recouped within a reasonable period.

In addition to the indications given above there are some other factors which may influence your decision. These can be summarised by the following questions:

1. *Do I have to use expensive staff for repetitive, routine operations?* If you do then it is worth while asking whether the process can be standardised and be written concisely enough to fit on to a microcomputer.
2. *Do I wish to standardise office procedure in certain areas?* If so, will the machine assist in educating staff into the new routine? It can do this by prompting the operator to input in a set way and by outputting information according to an office standard. In creative work, this form of standardisation can of course be a severe handicap. In any case this should not be a major reason for adopting a computer.
3. *Do I specialise in a particular building type?* Firms that specialise tend to have more data of a useful nature for that building type and can therefore build the models required for computer appraisal. The greater the degree of repetition of that building type the more useful and profitable the models will be. Successful models have already been written for appraising such diverse types as warehousing, hotels, offices and schools.
4. *Will the machine assist in public relations with my client?* At the present time it is an unfortunate fact that some clients will believe output from a computer in preference to an unsubstantiated statement from their professional advisers. The mystique of the machine will almost certainly disappear in coming years but a firm actively using a computer may well create a more 'go-ahead' impression in a competitive market.
5. *Do I anticipate a heavy involvement in computers in the future and therefore need to build up expertise at this early stage?* There is no doubt that computers are already an integral part of our life style. The pressure to use them is being built up with government encouragement in our schools and commercial concerns. An installation in the office may realise the expectations of school leavers and graduates thus encouraging the better recruitment of staff. In addition the installation will undoubtedly provide a better understanding of what is required in a larger installation to satisfy your more comprehensive computing needs at a later date. This expertise, which can only be learnt with 'hands-on' experience, takes time to develop. It may be worth investing in a cheaper machine at an early stage to avoid complications later on.

6. *Have other practices been able to usefully employ the computer for similar problems?* All things being equal if the answer is 'yes' then careful consideration should be given in case you lose your competitiveness.

7. *Is standard software available for my office techniques?* If it is then these programs will enable a rapid implementation of the machine and will save considerable sums of money in building up a suite of programs from scratch. It will also increase the credibility of the machine where there are sceptical office staff. As we shall see later it is the cost of developing software which is probably the most expensive item when installing a machine.

These questions are just a few of those that you may consider. In the end it comes down to, 'Do I have a problem?' and if so, 'Will a computer help me?'

Which system shall I buy?

The proliferation of equipment has made this question most difficult to answer. A check list of items is the best that can be offered in a book of this nature. The choice should be geared to the system or range of problems with which the machine will have to cope. Software systems are discussed in chapter 4 but suffice it to say here that an absolutely optimum fit of machine to current and future requirements is unlikely to be achieved. It would be rather like a child having an irregularly shaped solid and trying to pass it through the standard regular holes of a 'shape posting' cube. He would have to find the shaped hole that would allow the solid to pass with minimum gap in order to find the optimum solution. No exact fit will be possible and in some cases the task may be impossible and a larger cube with bigger holes will be required! The needs of the user cannot always be clearly defined and the most apt of the available machine configurations is therefore difficult to establish.

What can be said is that the choice of the machine should be software dependent. It is unwise to buy a machine and then see what you can do with it. As clear a view as possible of the system to be employed and its future development is essential if disappointment is not to be invited. Unfortunately with microcomputers, unlike the larger machines, the level of advice by the manufacturer as to which size would be most appropriate is minimal if not non-existent. This is the penalty to be paid for the low cost of these products. Wherever possible expert advice and preferably independent opinions from users of the proposed machine should be sought. Where a standard package of programs is to be used then the range of hardware options is likely to be very limited; if you will be developing your own software then the market is wide open. But it is important to realise that you may wish to market your own programs at a later date and therefore it will be an advantage to adopt a popular make of machine which is easily obtainable if considerable rewriting of programs to suit others is not to be required. Some of the major factors with regard to purchase are now discussed.

The central processing unit and keyboard

The CPU is at the heart of the hardware for the computer system. It is therefore of the utmost importance when considering purchase and will have an impact on all the other units in the system. The major items for consideration are given below.

Memory size

Although memory requirements depend upon the size of the program and system to be operated the minimum that should be purchased for any useful building application is 32K (approximately 32 thousand computer bytes) of available memory. Anything smaller will result in memory space problems for all but the simplest of applications. The word 'available' is included because some machines will use a part of their memory for the operating system and language conversion of the CPU. Thus some 48K machines will use 12K of their store for holding the BASIC language interpreter, leaving just 32K free for your program and the storing of data. In others the language is 'hardwired' into the machine either fixed or as a plug-in option (referred to as read-only memory or ROM) and almost the total capacity of the machine is available. Even with the latter type some of the memory may be used for machine operation and screen memory (i.e. temporarily holding what is presented on the screen) so that 1K (1024 bytes) or more of storage will be lost.

It is good policy if you are serious about developing a micro system (as opposed to using it as a stepping-stone to a mini) to buy as much memory capacity as you can. It is usually cheaper to buy the increased memory with the initial purchase than to try and upgrade at a later date. 128K micros are now available but the majority of current micros have 16 wires on what is called the 'address bus' and this imposes an absolute maximum of 64K (2^{16} = 65 536 = 64K) words of memory for direct addressing. Main core memory is faster, simpler and more efficient than using backing store memory and this should be a major consideration when developing large programs or programs which generate large amounts of intermediate data.

There are rules of thumb for gauging the length of program that can be handled by any particular machine. These are based upon assumptions regarding the number of bytes of information per line and the amount of information to be generated and held within the machine. However 'typical' programs of this type are seldom wanted. Building progams which are not of the simple 'solve a formula' type usually require files of data (e.g. on performance and cost of components) to be held and to be accessed by the machine. They may also require large tables or matrices to be held and operated upon. Both these demands add extensively to the memory store requirements. In fact where large quantities of data are to be used in conjunction with a microcomputer it is almost invariably necessary to use disks and to bring the data into the main memory in chunks only when it is required. This complicates the programming and adds to the processing time. If the kind of application you have in mind is largely of this type then serious consideration should be given to the purchase of a minicomputer.

To give some idea of the amount that can be stored in memory it may be helpful to consider how many pages of the *Concise Oxford Dictionary* can be stored in various types of memory. If we assume 1 byte per character it works out at approximately 5 kbytes per page and about 5 Mbytes (megabytes) per dictionary. We can then produce table 3.1 for comparative storage.

Table 3.1 Storage capacity for *Concise Oxford Dictionary* (based on *Microprocessors — A Short Introduction*, Department of Industry).

Storage	Capacity (in dictionary pages)
32K microcomputer (available memory allowing for some machine housekeeping)	6 pages
5 inch diameter floppy disk (capacity of 200K or 0.2M. This can be doubled using double-sided disks)	40 pages
8 inch diameter floppy disks (capacity of 1000K or 1M)	200 pages
C60 cassette tape (capacity 400K or 0.4M)	80 pages
C15 cassette tape (capacity 100K or 0.1M)	20 pages
15 inch diameter rigid disc (capacity 10M)	2 dictionaries

Another way of looking at it is to say that the complete job records required for formula price adjustment using the NEDO formula on a twenty-four month project would take about 20 to 25K of disk space. The program to access that information every month and to carry out the calculation would take about the same amount of main core memory although in both cases it will depend upon the sophistication of the routines.

The simple question to ask is, 'What is likely to be the limit of my use of the microcomputer?', and then to buy a machine with more than enough capacity and with a safety factor of at least 50%. Very few people have bought microcomputers and found that they have wasted memory but unfortunately the reverse has occurred all too often. The rule of thumb is, 'Do not buy at the top of a cheap range, even if it appears to overprovide for your requirements; buy a machine at the bottom of a more expensive range which can expand'.

Speed and ease of operation

The implications of computer 'word' length and speed of operation have already been discussed in chapter 2. Most of the popular micro-computers use the 8-bit word (the equivalent of 1 byte). This is satisfactory for a wide range of applications but because the amount of information executed in each operation is fairly small the processing time will be slower. If you are involved in heavy 'number cruncing' routines, major searches through banks of data or large simulation packages, then a 16-bit word processor will be essential. Some of the more expensive micros have this facility but you will probably need a mini.

Good programming is likely to have the greatest effect on speed of operation. If your programmer can work in 'machine code' (i.e. not in a high level language) then the time saved in compiling or interpreting the program will result in much faster processing. Even using a high level language (and some are faster than others) it is possible for a good progammer to organise his routines and subroutines in such a way as to make the program run efficiently and therefore faster.

The 'translator' (i.e. the program that translates from a high level language to machine code) will also have an effect on speed. If the computer uses a compiler, where the whole of the 'source program' (original) is translated into machine code before execution by the computer, then operation will tend to be faster. The penalty how-ever is that these programs use a lot of memory space and there-fore limit the extent of application of microcomputers. 'Interpreters' on the other hand translate each program statement just before it is executed. These programs are easier to write and more economical in memory and are more commonly used in micros. Their disadvantage is in speed of operation which can be ten times slower than compiled ones. This may not be critical on an in-house machine where maximum utilisation is not required.

The major difference in machines, the characteristic which decides the personality of each one, is due mainly to the in-built sophistication of the ROM operating system and translating program. In short, the program already inside the machine when you buy it will determine how pleasant it will be to write or operate your own programs. In crude terms the larger the ROM the more 'friendly' the computer will be. In some models there may be a few empty sockets on the circuit board enabling the owner to insert extra ROMs. These can be very useful for using preprogrammed ROMs such as a programmer's toolkit, high resolution graphics or a method of organising and manipulating data. The effect of these 'add-ons' is to improve the power and versatility of the machine and is almost equivalent to a normal upgrade or enhancement of memory.

Languages
If you are not familiar with computers then the range of languages available for programming will appear confusing. Nearly all micro-computers will however accommodate the language of BASIC. It is a simple, easy to use method of giving the machine instructions which corresponds closely to written English. Since its original inception it has been developed considerably by the manufacturers to compare more with the older program languages such as FORTRAN. There is not an absolutely standard set of agreed instructions and therefore a few unique statements will appear on every machine, but these are easily absorbed.

The adoption of BASIC by most manufacturers has encouraged the writing of software packages in this language and allows the rapid transfer of programs between machines. Unfortunately this cannot be done by automatic transfer (e.g. a cartridge tape) because there is seldom compatability between machines of different makes. The program has therefore to be slightly altered to suit the new machine and then input via the keyboard and recorded in that machine's backing store.

Some of the popular micros allow other languages to be used merely by plugging in a new ROM containing the new language. In other cases where the language is disk based it is just a question of entering the appropriate language disk into the memory before keying in the program. This disk type does of course use up some of the main core memory and therefore can substantially reduce the amount of memory available for programming.

Many of the early programs written for building applications were written in FORTRAN and indeed it is still one of the most popular languages for large construction programs today. As most of these programs are too large to run on a micro the need to use FORTRAN is not very great. However many programmers still prefer to use this language as it still has some advantages over BASIC. Most will adapt to BASIC very quickly however and this should not be a major problem.

For the programs in sections II and III of this book we have kept to BASIC because of its almost universal application to small micros. It has been suggested that in the medium term future another language, PASCAL, will overtake the popularity of BASIC, but we can only wait and see.

Keyboard

Although strictly a peripheral device the input keyboard is included in this section as it often comes as an integral part of the CPU container. Now an important consideration in buying a computer is, 'How easy is it to operate?' We have already mentioned how the operating system can make the machine more 'user-friendly'. Fortunately, the complex mystique of the main frame has disappeared and initial appearances are not as off-putting as they once were. Most of the commercial type of micros have a standard QWERTY typewriter keyboard and occasionally a separate number pad (i.e. a group of keys for inputting figures). This layout has the advantage that a typist can rapidly enter information, including the program and data. Micros have been developed to be used largely in 'interactive mode'. This term refers to the keying-in of information in response to a question or command from the video monitor when the program has paused in mid-operation. Once the machine has received the data it will continue to progress through its instructions using the information given until it receives another instruction to pause. It is sometimes called 'conversational mode' because it is similar to holding a very stilted conversation. Easy use of the keyboard is thus essential particularly if large amounts of data are to be input. The layouts found on some of the very small machines which require excessive use of the 'shift' key or have the keys very close together can be very irritating and cause errors. Checks that should be made are:

(i)   Has it a QWERTY typewriter layout for speed of operation?
(ii) Has it a separate number pad for fast input of figures?
(iii) Are the keys positioned sensibly and of reasonable size for rapid fingerwork (e.g. typewriter size)?

Standard functions

Most micro keyboards will have additional features over and above that of the normal typewriter. Special instructions, cursor movements (i.e. movement of the position of characters on the screen), graphics and special signs will often be found. In the case of instructions, for example to clear the screen, this can speed up the keying in process and very often will save on memory.

Although we shall not refer to graphics too much in this book this facility can be extremely useful. Where the graphics character is input via the keyboard in the same way as a figure or letter then it is almost certainly 'low resolution' graphics. This means that the character fills one whole space reserved for a character on the screen. It cannot be enlarged or reduced. Therefore a graph plot on a 40×25 character screen would be limited to that number of positions (i.e. 1000). In 'high resolution' graphics each character space can be divided into a much smaller number of units known as 'pixels', often at the discretion of the user, and therefore more precision in plotting can be achieved. At the same time a penalty has to be paid in terms of memory store used for this facility. A 160×256 plotting points matrix may require up to 20K of RAM. For everyday work graphic symbols can be of great help in setting up interesting, even animated, titles to programs and for drawing simple details, say for building measurement, histograms, or to aid a description when considering component performance.

Machine manuals

A short word on manuals would not go amiss at this point. They have an important impact on the way you approach the machine and the way you learn and develop your skills in conjunction with it. A poor machine manual will result in many hours of frustration and impatience at your speed of progress. The more popular micro manufacturers have realised their early errors in this respect and now provide very useful self-teaching guides. Some of the personal computing journals also provide extra guidance and 'program tricks' for the popular makes. If you decide to choose a 'minority' machine then ask to see the manuals before committing yourself particularly with regard to disk operation and file handling.

Upgrading and compatability

In view of the fact that you may not want the all-singing all-dancing version of your micro to begin with, there are some important questions that should be asked, and satisfactory answers received before purchase.

(i) Can I extend the memory store at a later date and how much will it cost?

(ii) Will my current programs be compatible with any further hardware development of this machine?

(iii) Can I access any other machines with this micro? Salesmen tend to be rather vague on this point and full details should be obtained. Standard 'interfaces' between micros and mini machines are sometimes available and some machines are better than others. Do not rely on the casual comment that 'Professor Bloggs at Cambridge has managed to do it'. Certain engineers have the ability to get the apparently impossible working. It will usually be beyond the ability of us poor mortals unless it is simply a question of plugging two off-the-shelf bits of equipment together. Where this is possible the machine can become very powerful indeed. If you are thinking of upgrading to a mini at a later date then careful consideration should be given to future compatability. The CP/M (control program/monitor) operating system now available on most microcomputers has increased the number of minicomputers that can be accessed providing they, too, operate this system.

(iv) Can I access a standard data base? With the increasing development of the television data bases such as 'Viewdata' and 'Prestel' it will be helpful to be able to access this screen information and to use it in your programs. Some of the small popular micros now include a plug-in board which will allow this type of data to be input to the machine. Quantity surveyors will be able to access the UK RICS Building Cost Information Service and financial statistics. At the moment there is little such information for architects.

(v) Is there standard software available for this machine that I can make the basis of a personal system? This will be discussed later but a lot of hard work may be saved by building up your own system around one of the standard pieces of software available from the major manufacturers. This particularly applies to data bases but may also include statistical routines and programs for scheduling. It will be necessary to check that other programs can be 'bolted on' to the standard software without too much trouble.

Number of peripheral devices

Peripheral devices (e.g. disk drives, printer, plotter etc.) are discussed later. Suffice it to say here that there should be ample scope for adding on all the bits and pieces you will eventually require to make your complete system. Some of the larger micros will allow more than one keyboard to be run off the same central processing unit. This may be of great assistance where several members of the firm are likely to be using the same program. Even with the smaller machines, 'multiplexers', as they are called, are sometimes available for the same purpose. The problem of course is the limited amount of memory available when the memory core is divided up for each user. Nowadays there is unlikely to be a problem with peripherals unless you want to run something like a printer and a plotter at the same time!

Maintenance

As with all mechanical/electrical devices computers can sometimes go wrong. Occasionally a chip will fail or its connection become dislodged. It is vital in a commercial application that servicing and maintenance of machinery can take place rapidly. Questions should be asked about availability of spares, nearest stockist and engineer. In addition the response time for an engineer when under a maintenance contract should be ascertained and if possible some guarantee be given in the contract itself. As the machines are so portable some suppliers will provide a replacement machine for use until the old one has been repaired. Maintenance agreements are likely to cost between 10 and 15% of the purchase price and you may feel that with a small system that does not involve expensive mechanical peripherals then a maintenance contract is inappropriate. It will rather depend on:

(i) How important is it for you to have continuous operation i.e. immediate repair? A maintenance contractor may well promise a 24-hour service.
(ii) How good a service can your local supplier provide, without a contract?
(iii) How popular is your machine? Theoretically the more popular the make the wider the range of spares held and, probably more important, the greater the degree of expertise available for diagnosing and rectifying faults. Both these facts will influence your decision as to whether you require the security of a maintenance contract.

In the authors' experience the smaller popular microcomputers, with a good supplier in close proximity, probably do not require a maintenance contract. With the more sophisticated machines, purchase price say over £5000, the balance changes to the other direction.

The peripherals

No central processing unit is of much value in its own right. Information must be input and output to derive benefit. The parts of the system which connect to the CPU to allow this flow of data are called the peripherals and are usually plugged in to 'ports' on the side of the main unit. Since the building professional is going to spend much of his time feeding facts and figures into the machine and will expect good recording and presentation of output, these items of equipment can prove to be just as important as the CPU. There is no doubt that good presentation of results in particular will have a marked effect on the confidence of the user and his client and sustain the credibility of the computer.

The complete range of peripherals, with their pros and cons, is large. It is therefore the intention to concentrate in this book on the more commonly used items in microcomputer systems.

Backing store

Backing store or mass storage memory is the cheaper method of extending the main core memory and at the same time providing the opportunity for making a permanent record of the program and its data. It is different from main memory in that it is not directly accessible to the CPU. In fact 'chunks' of information are copied into the computer's memory, used and possibly adjusted, before returning for storage purposes. This copying process via an electromechanical device is slow compared to main memory which has no moving parts. However since it can record information on a magnetic medium it can be used to hold banks of data and programs for use in a complete computer system. The recorded parts, tape or disk, are limited in size, but by interchanging them an inexhaustible supply of information can be accessed.

CASSETTE TAPES

When micros first came on the market the most popular way of creating 'mass storage' was by means of a cassette tape. The blank tape is identical to those used for the reproduction of music or speech and is used in a recorder that is compatible with the machine. The program held in computer memory, that would have been lost if the machine had been switched off, can be recorded and then fed into the machine when next required. The advantage of the tape is its portability, cheapness and ruggedness. Its major disadvantage is the time taken to record and input messages. This can be several minutes for a long program compared with a few seconds on a disk. Although the tape can be used for holding file data which can be input to or output from the computer at a relevant point in the program, the time constraint generally makes this impractical. The overall speed of the tape is also reduced by the fact that any item of data stored as a tape must be accessed 'sequentially' (i.e. the tape must be played through until the item needed is found). With a C60 tape this could mean up to half an hour waiting time!

Some experimentation has been undertaken with 'floppy tapes' which work on a continuous loop principle and at a much faster rate than conventional tapes. They need a special record/play unit within which to operate. At between one quarter and one third of a disk price they will speed up the operation of a tape system by about 5 to 8 times depending on the program. Whether these will capture enough of the market for them to become a popular choice remains to be seen. It would appear that some problems with regard to reliability also need to be overcome.

A recurring problem with all small micros is the lack of standardisation which will stop a tape or disk being used on another manufacturer's equipment. For example until recently a program recorded on a PET cannot easily be transferred to an Apple (despite the fact that they use the same microchip) without keying the whole thing in again and possibly having to alter some program lines.

Some devices have been produced which will allow the 'interfacing' of these smaller machines but the degree to which this can be done will need to be investigated. Not all commands will be compatible, for instance. At the slightly more sophisticated micro level such as the North Star Horizon and Superbrain there is a higher level of compatibility between the machines because they tend to use the same operating system and the same microprocessor, i.e. CP/M with the Z80A chip. CP/M was first developed in 1973 by Gary Kildall. It is a disk operating system for microcomputers, produced by a company named Digital Research, and was designed for use on 8080- and Z80-based microcomputer systems. CP/M stands for 'control program/monitor' and versions are available for a wide variety of microcomputers using 8 inch and 5¼ inch floppy disk drives. CP/M is fast becoming the industry standard and both PET and Apple now allow for use of this system.

A particular difficulty relating to some of the inexpensive tape players to be used in conjunction with the machines is that occasionally the recording heads do not line up. Consequently when using different tape players of the same type and with the same computer, the machine will not transfer the program. This can be overcome by realigning the heads to make all the machines compatible but it is worth checking for compatibility if a series of independent machines are to be installed in a single practice.

FLOPPY DISKS

These are the most popular method of providing backing store for small micros. The disk unit comprises a motor for rotating the disk and a magnetic head which reads from the disk, housed in a casing which generally takes between 2 and 4 of these drives. The disk itself looks like a flexibile gramophone record and is protected from damage in handling by a square paper sleeve. The area of the disk is divided into tracks and sectors so that areas of it can be separately addressable. An item of data can be accessed by the magnetic head moving to the specified track and reading the required item when the disk, which is spinning rapidly, is at the right position for the appropriate sector (see fig. 3.1).

The capacity of a disk depends on the density with which information is packed and whether it is single or double sided. In addition there are two sizes of disk, 5.5 inch and 8 inch. A single-sided, single-density 5.5 inch disk will hold between 80 and 250K of information but if this was raised to double density then as expected the amount of information also doubles. Likewise a double-sided disk would increase the amount of information by a factor of two. The larger 8 inch disk will hold between 250 and 300K of data for a single-sided, single-density disk with a similar relationship existing for double-density and double-sided disks. The variety of formats and densities for disks makes it seldom possible to use the data prepared on one computer on another of a different make unless they use CP/M.

A point that should be made here is that at least two disk drives should be provided. This is to allow rapid copying of information from one disk to another to ensure that a duplicate is kept for security reasons. It also helps in the inputting and outputting of data on separate disks for future use.

The advantages of floppy disks are that they are cheap, easily stored, convenient and easy to use. Their disadvantage is that they are neither as reliable nor as fast as hard disks. Particular care must be taken to ensure that the disks are not abrased in any way or, as has happened in the past, used as a coffee cup mat!

With disk systems in particular it is vital to get good information on performance, preferably from another user with the same requirements as yourself, before buying. Access times can vary by a factor of 3 and the ease with which 'random access' can be achieved varies considerably. Random access is the ability to go straight to an item of data in a file instead of reading sequentially through all the file data to find it. This form of access is essential for speed of operation and also the speed at which individual pieces of data can be updated. It is probably worth paying for tuition on disk operation unless you can persuade your dealer to include this as part of the purchase price.

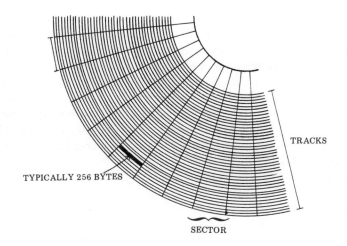

TRACKS

TYPICALLY 256 BYTES

SECTOR

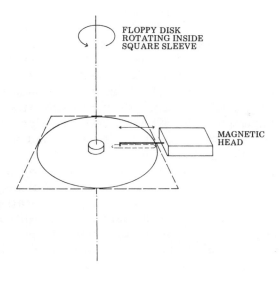

FLOPPY DISK
ROTATING INSIDE
SQUARE SLEEVE

MAGNETIC
HEAD

Fig. 3.1 Floppy disk.

HARD DISKS

A type of hard disk is available which was specially developed for use with small computer systems. It is known as the 'Winchester' disk, is cheaper than other hard disks and can have a capacity of 15 million characters (15 Mbytes). It is also much faster and more reliable than the conventional 'floppies' but has some disadvantages. It is more expensive than a floppy disk and is completely sealed; this means that copies of the stored information, necessary for security purposes, can only be made by transferring the information on to tapes or floppy disks. A second 'Winchester' could also be used but this tends to be an expensive solution. There is no doubt that engineers are working on this problem and this particular difficulty should soon be overcome.

BUBBLE MEMORIES

This kind of memory fills the gap between the fast but volatile semiconductor memory that is found in the computer main store and the low cost but slow and sometimes unreliable floppy disks and tapes. It uses small cylindrical magnetic domains or 'bubbles' in a thin film of special garnet crystal. The domains are magnetically polarised in the opposite direction to the remainder of the film. The magnetic bubbles can be moved about the film by means of electrical pulses applied to the bubble memory connections. The presence of a bubble corresponds to the binary digit 1 and the absence of a bubble at a certain point to a binary digit 0. Since the 'bubbles' stay in position when the power is switched off the information is held ready for use when it comes on again. Memory access times are up to 100 times faster than disk although remaining slower than main core memory. Bubble memories of 1M are currently available but their cost is much higher than conventional methods of storage. However it has been forecast that the bubble memory market will be worth $1000 million a year by the end of the 1980s and therefore there is considerable investment in their development. An introduction of bubble memory on this scale would have a major impact on the microcomputer market and its potential. At the moment we can only wait and see.

Printers

The purpose of the printer is to take a piece of information from the computer and record it on paper. This can be done in broadly two ways. One category would be labelled 'impact' printing and the other 'non-impact'.

Under impact printers we can include 'dot-matrix', 'daisywheel' and 'golfball' types. A matrix printer creates its characters by using a set of needles. The number of needles used will affect the quality of the character produced. For example, creating the character out of a 5 × 7 head (a common size for cheap printers) will not be as good as that produced by a 9 × 7 head. Daisywheel and golfball printers produce a much higher quality type in a manner similar to that used by a conventional typewriter. They are normally used for correspondence and reports and usually entail between three and four times the cost of a matrix printer.

The speed of the printer is fairly critical where long lists of results are expected. A printer operating at 30 characters per second (cps) is considered to be low speed for a dot-matrix printer and 150 cps is medium speed. Daisywheel machines are likely to be at the lower end of this range. The high speed line printers associated with larger minicomputers are not generally used with the type of machine considered in this book.

The non-impact printers use special paper which is sensitive to heat or electric charge, tends to be expensive, and does not always photocopy well. Heat sensitive paper also tends to fade in prolonged exposure to sunlight.

Points to query when selecting a printer are:

(i)   How fast does it print?
(ii)   Will it successfully interface with my machine without any loss of speed or performance?
(iii)   What type of paper feed does it use? (NB Tractor feed is much more suitable with continuous stationery.)
(iv)   Does it need special, expensive paper?
(v)   Does it allow different sizes of type or enhancement of the characters?
(vi)   How many columns will it print across the page?
(vii)   Does it allow me to print the graphic symbols of my machine?
(viii) Does it allow me to plot graphs and histograms?
(ix)   How noisy is it when operating?

Note that if the printer is bi-directional it means that it prints alternate lines backwards obviating the delay of return to the left hand margin. This speeds up printing and therefore a 120 cps bi-directional printer will in fact be faster than a similar speed mono-directional printer.

Visual display units (VDUs)
Once again the range of possibilities here is considerable although most microcomputers either have a built-in screen display or have a recommended television monitor. The best type of screen display is a monitor specially designed for computer work. It has a more

stable screen image and tends to use a relaxing colour for the characters, avoiding glare and eye strain. It may not seem important initially but many hours will be spent in front of the screen and this can be agony if the image flickers or is difficult to read.

The VDU saves considerable wastage of paper as all intermediate questions and results can be output through this peripheral rather than through the printer. Only those results which are required as hardcopy need be printed. In those programs which are predominantly text oriented the number of characters per line must be a major factor in your choice. Many microcomputers only manage 40 characters per line which is a pity; the restriction imposed by this number of characters becomes particularly apparent when displaying information in tabular form — there never seem to be enough room to squeeze in the columns needed! 80 characters is a much more attractive proposition provided that the characters maintain their definition and clarity. The number of lines per screen page should ideally be 50 but 25 is more common.

Some computers offer colour, i.e. it is possible under program control to present the text in various colours. It is unlikely that this requirement will be a necessity for building appraisal programs but there is no doubt that the ability to intersperse various colours in diagrams or pictures adds an entirely novel dimension to computing. This luxury does however mean not only some special circuitry in the computer itself but also the expense of a colour television.

Although with many machines it is possible to use a conventional television set, it is worth noting that the UHF modulator in the set will probably require the computer to circumnavigate some complex tuning circuits before it arrives at its screen destination. Unfortunately this process does sometimes distort the signal sufficiently to produce a certain amount of 'fuzz' around the edge of the characters. It is better to purchase a proper computer monitor which contains no high frequency tuning circuitry and in which the signals from the computer are clearer and fed directly to the video amplifier and deflection controls of the tube. If you are contemplating high resolution colour graphics then a colour monitor does become almost an essential since a domestic colour TV cannot cope with the high switching speeds that the computer produces via the modulator.

A final word; although there appears to be no firm evidence of radiation hazard, it is advisable to sit well back from the screen merely to avoid eye strain.

Miscellaneous peripheral equipment
In addition to the basic units of a typical microcomputer system outlined above there are any number of other 'goodies' which can be

attached. They vary from very useful plotters and digitisers to items which although possibly helpful can really be described as novelties. The latter include audio devices which play sounds to order; oral input devices which recognise (after tuning) the human voice and translate the sounds into characters; 'scratch pads' which allow characters written by hand to be recognised by the machine; and 'joy sticks' for the movement of characters around the screen. With refinement there is considerable potential for these devices particularly those concerned with oral input. One of the problems we face at present in computing is the time consuming task of typing in and checking data. How nice it would be if the building economist could dictate all his measurements to the machine together with a description code, press a button and have printed out a bill of quantities together with a copy of his dimensions. Or, for an architect to call up a component verbally and be given its performance characteristics, or access to a standard specification item for the preparation of contract documentation. These devices are not yet sufficiently reliable to undertake these tasks; researchers are working in this area but it will be some time before a system is developed in which absolute confidence can be placed.

Plotters and digitisers are of course already with us. Whether a really satisfactory system including these two peripherals can be achieved with the limited capacity of current microcomputers remains open to debate — where the peripheral itself contains some extra memory store as well as the operating system it may be possible. There are however specialist items of equipment and a careful specification would need to be drafted for any particular need. Special care should be given to the accuracy of measurement from the digitiser and the accuracy of plot on the printer. Speed of operation, flexibility, size of working surface and reliability should all be determined and it would be wise to ask for not only a demonstration but also loan of the equipment for a reasonable period before deciding to purchase. They are expensive pieces of equipment and it is essential that you ascertain their potential and limitations before making a commitment if disappointment is not to be risked. A detailed consideration of these items is outside the scope of this book.

Calculating the cost

Inevitably the final selection of which machine is to be purchased will depend mainly on how much money is available. Of course purchase is not the only option as many machines can be rented or leased, but most firms providing this facility require a pay-back period of less than a year. Since it takes several months to get a system fully under way and integrated into the office routine, it is unlikely that you will want to change your machine in such a short space of time. Unless you wish to try out a number of machines over a short period for comparison purposes then renting or leasing is unlikely to prove the right option despite some marginal tax advantage in the UK. If you do decide to venture into the minicomputer or upper end of the microcomputer market then things may be different. The leasing or rental agreement will almost certainly include maintenance of the equipment and replacement of defective equipment; other advantages are that you can upgrade to the new improved version of your machine when it becomes available and large capital sums are not required as the rental can be paid out of revenue. However it is unlikely that the circumstances of a professional office are such that purchase proves to be other than the most attractive proposition.

What then are the costs involved in the purchase of a typical microcomputer system? They can be classified under five headings.

INITIAL COST

This is the cost of purchasing the computer and its peripherals. A small system comprising CPU, visual display unit, disk storage and printer will cost between £2000 and £5000. There are generally no installation costs and a manual is nearly always provided free. In addition a good supplier will give some preliminary instruction. This is in his interest otherwise you will be plaguing him with questions when you do not understand why the machine may not be functioning as expected. This initial outlay is currently subject to tax relief in the UK.

Typical cost:

CPU	1250
VDU	700
Disk drive	600
Matrix printer	450
Software package etc.	550
Total cost	£3550

A first year tax allowance of 100% of the cost may currently be claimed in the UK and is given by simply deducting the full amount from profits before calculating the amount of tax payable. The value of this in saved tax will obviously depend on the rate of tax suffered on partnership profits.

Thus, if the profits are low enough to be taxed at 30% standard rate the relief is worth 30% × £3550, i.e. £1065, giving a net cost of £2485.

If the maximum rate of tax is 55% then the relief is worth 55% × £3550 (£1952.50) and the effective net cost comes down to £1597.50.

MAINTENANCE COSTS

We have already suggested in this chapter that a maintenance contract on a small system may be an unnecessary expense. It will probably cost between 10% and 15% of the initial cost per annum and it may be difficult to argue that this is justified if you have a popular machine with a good supplier nearby. Loose connections and teething troubles are likely to occur in the guarantee period. Large scale replacement of internal circuitry is unlikely and is in any case relatively inexpensive. Disks probably cause the most trouble but repair is not usually that expensive. The advantage of the maintenance contract is the speed at which the problem is rectified. If the service can be guaranteed within say 24 or 48 hours then it may be worth this extra operating expense to keep the equipment in constant use. It should be noted that rough handling of the equipment is likely to lead to increased maintenance; too many journeys on the back seat of a car or shifting around the office is likely to lead to more problems than a fixed station in a quiet room.

CONSUMABLE COSTS

There is obviously a relatively small amount of money to be spent on such things as paper, disks and tapes. However with disks currently costing about £3 to £5 each and 'continuous paper' about £15 a ream the extra cost is not a great deal more than the paper and files plus storage space that has been replaced.

SOFTWARE

Of all the items this has the potential to be the most expensive; in fact it may far outweigh the cost of the original hardware. This may not be self-evident on first viewing as it will be possible to buy off-the-shelf packages for energy evaluation, NEDO formula price adjustment, daylight calculations and so on for between £100 and £300 each. While these programs will no doubt work very well there will come a time when you want either to adjust the program to suit your own particular requirements or to write a completely new program for an application suited to your office. This is when it can start to become expensive. Writing programs is a time consuming business. Not only does the application have to be carefully analysed and any problems solved but a great deal of structuring is required for the program to be used as painlessly and efficiently as possible. It may take a week to write a program that will successfully perform the task but it will take probably twice that long to make it into an 'idiot-proof' and operational package. Much of the time will be spent just correcting mistakes both in typing and logic — this is known as 'de-bugging' and can be the most frustrating part of computer work. Providing foolproof checking routines for user input is also time consuming although an experienced programmer will build up a library of routines to cope with these eventualities. In fact the experience of the programmer will show itself in the number of standard routines he has developed, his efficiency in writing the program with optimum speed of computer operation, and his grasp of the exercise in hand. For a small modest program it will probably take a month to provide a package that is usable in the office and as it is used so new refinements will be demanded and will consume further time.

It is impossible to be too specific as to how long it will take to develop a program but fig. 3.2 may give an indication of the proportion of time spent on each aspect of software development.

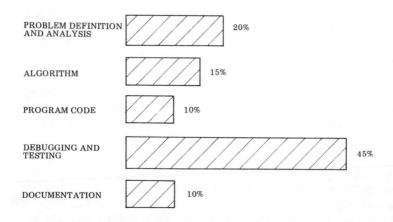

Fig. 3.2 Typical proportional breakdown of time related to the development of a program.

It is the view of the authors that if the firm is large and serious about computing then it pays to employ a programmer/systems-analyst who can not only write the programs and the description of their operation but also educate the staff into the use of the machine. Attempting to do this with existing professional office staff on a part-time basis seldom proves satisfactory, however enthusiastic and keen the staff are. Another option is to employ a software house to write the programs but this tends to be fairly expensive because of their overheads and it is also rather more difficult to refine the program once it is in operation. An in-staff programmer also tends to get to know the office practice, technical vocabulary and personal whims of staff (an important consideration in this respect is the office documentation for the program). It is not until the manual has been followed by the 'lay' staff in the office that its deficiencies begin to appear and it is therefore useful to have someone available who can make improvements.

Another problem in using a software house for the kind of small systems to be used on a micro is that it is sometimes difficult to define with clarity your precise requirements for the program. Too much changing of your mind or refinement could take you outside the original terms of the contract and into further expense.

Having made these points, there is no doubt that in the majority of computer-oriented firms a keen member of the professional staff is at the root of the development of programs for that office. Although the systems developed often tend to be an *ad hoc* grouping of programs and in some cases written without optimum efficiency, the end result is a series of useful programs which are suited to the needs of the professional architect or surveyor. The decision to be made is whether the firm can afford the expense of that person's time to provide that kind of development, with its advantage that at the end of the day the building professional will be fairly heavily committed to programs of his own creation, and that his own enthusiasm is likely to encourage their use in the office.

HIDDEN COSTS

These are the costs which may not always be anticipated and yet should really be charged to the implementation of any computer system. An obvious one is the extra cost of insurance for accidental damage or loss of the machine. This is a fairly small item on the list of charges but is well worth while.

The major hidden costs, however, relate to the accommodation for the machine and the cost of implementing the machine in the office.

Accommodation is not so much a problem as it once was. Gone are the days when air conditioning was required and temperature control was essential. Most micros are fairly tolerant of environment and now it is the operator who is more likely to complain about his or her conditions! Additional office space, desk and possibly a 'clean' electricity supply (to overcome possible distortions which may upset certain machines) may need to be provided. Desks of a more convenient shape for computing are available but tend to be expensive when compared with conventional furniture which is often satisfactory. Siting of the machine can be fairly critical to avoid undue glare or reflectance from the screen.

The major hidden cost is that of implementing the machine into office practice. Under this heading can be included:

(a)   The cost of setting up the machine — not usually a very expensive item.

(b)   The cost of training personnel to use the standard programs. This can be quite expensive, particularly if the program is to be used by the whole office. Staff will need to know the rules for switching on and off, the commands for operation, the alternative outputs and routines that are available and so on. This will not be just the cost of a short course but will include a period of 'playing' with the machine and program to gain confidence in its operation.

(c)   The cost of training the professional staff to program the machine. Even if staff will not be writing the major software it is of great help if some of them at least know the principles of how to program. This knowledge will enable them to have a better understanding of the capabilities and limitations of the machine; perhaps more importantly they will also gain knowledge of its potential, and how to structure applications for the briefing of a specialist programmer, appreciate the difficulties of his task and appraise the extent of his achievement.

(d)   The cost of preparing a standard office practice for use with the programs and of course the handbook for each program's operation.

(e)   The cost of a partner's time in supervising the whole operation.

If all these extras are added up they can amount to a substantial sum of money and even exceed the hardware costs.

Computing the overall cost

Now let us look at the total cost of installing and operating a small microcomputer system. A small machine will probably have to justify itself over a period of three years because of the fast rate of wear of its mechanical parts. We will assume the following:

Initial cost	£5000
Maintenance	15%
Annual drop in residual value	50%

Software 'off the peg'	£2000
Annual costs	
Consumables	£200
Insurance	20
Staff training (year 0)	£500
Staff training (other years)	£250
Office reorganisation (year 0)	£250 and then £50 p.a.
Machine supervision (all years)	£500
Furniture etc. (year 0)	£300

The above items are conservatively priced and in particular it is assumed that software is bought in as an off-the-shelf package. These costs now need to be tabulated and discounted to find the real cost to the organisation (table 3.2).

Table 3.2 Example of costs for installing and operating a small microcomputer system.

Year	Capital cost	Annual costs	Residual value	Discount factor @ 10%	Discounted expenditure
0	5000	200			8 770
	2000	20		1	
	300	500			
		250			
		500			
1		750			
		200			
		20		0.91	1 611
		250			
		50			
		500			
		1770			
2		1770		0.85	1 469
3		1770		0.75	1 328
				Total	13 178
Less residual value			625	0.75	469
				Total	12 709
				Say	£12 750

Tax relief has not been included.
Assume no maintenance in year 0 as the machine will be covered by guarantee.

The final total in table 3.2 represents something of the order of £260 per month or the equivalent of about three man-days, meaning that at least three man-days should be saved per month by using the computer if it is to prove a satisfactory financial acquisition. Tax has not been deducted from the above but is unlikely to have a great effect as it is also deductible for the labour it replaces.

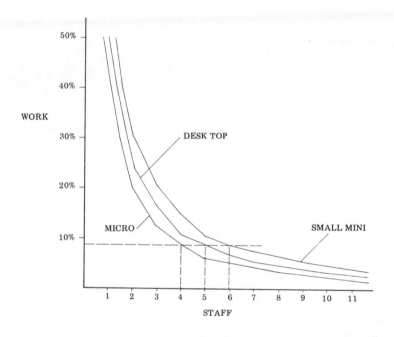

Fig. 3.3 Proportion of work required to be put into a computer in order to justify purchase related to number of staff employed. (Reproduced by kind permission of the Construction Industry Computing Association from their book *Micros in Construction*).

If the software purchased provides a saving of say 2 hours per month per contract (not uncommon for the NEDO formula) then 15 contracts would be required to break even. However it is most unlikely that the machine would be used for one type of program alone; indeed the major benefit of the machine is not necessarily to speed up an existing process but to do things which it would have been impossible to attempt by manual means. For example, if we can consider a much larger number of design options in the same amount of time, the chances of providing a better final solution are probably increased. This may not have an immediate visible pay-off because it is unlikely to affect the fee paid by the client, but in terms of the goodwill created by the design and the concern for the clients well-being the return will be seen in future commissions.

It is interesting to note that the UK Construction Industry Computing Association (CICA) in their publication *Micros in Construction* suggested that, 'assuming 10% of the work in an office is computable, then the cheap computers produced by the new technology are relevant to, typically, between 40% and 70% of all construction industry firms'. The amount of work required to be placed on a machine to make it pay will depend on the number of staff employed. More staff will provide more utilisation of the equipment. Fig. 3.3 gives a guide as to the relationship between work that is computable and the number of staff employed in order to make a certain type of machine pay.

Only broad generalisations have been given and each case should be considered on its merits. Once the machine appears to be a viable proposition then careful consideration should be given to the manner in which it is introduced into the office.

Machine implementation

Too many offices have foundered, not on the choice of machine or the suitability of the software, but on the way it is introduced into the office. In some cases the computer has been found locked away in the partner's office. On other occasions staff are allowed to see the machine and use it but 'any "development" must be undertaken in your own time'. Any new approach to professional practice will produce a reaction from staff who are often under pressure of workload and feel that they need to be fully convinced that the innovation will not be just a flash in the pan or time consuming novelty. If they are to subscribe their time to learning and mastering the machine they must feel it is worth while, and probably more important, that the firm is prepared to support them in their endeavours. The following suggestions are put forward as ways in which the management of the firm could contribute:

(i) Place the machine in a central position available to all who are potential users.

(ii) Provide regular instruction and seminars — not only at the time of introduction but at regular periods thereafter to maintain interest, especially when new software has been written.

(iii) Encourage development by staff, even if the programs are not efficient, as self-discovery not only will be the best way to learn but also will maintain motivation.

(iv) If possible allow the machine to be taken home at weekends for staff to develop programs (even if it will be partly used by the children eager for the latest game) as it creates a homely and personal image. Check first of course that the machine is sufficiently robust to travel well.

(v) When the machine arrives make sure there are some instantly useable and relevant programs available, so that the credibility of the machine is not lost while it remains inert in the corner waiting for someone to develop its use.

(vi) Appoint one person, preferably a partner, to be responsible for its development within the firm. This should be more than a token gesture — this person should be able to understand the machine and operate it as well as administer the system improvement.

(vii) Train one of the office or junior professional staff to look after the routine maintenance and instruct one of the firm's typists on input procedure so that long programs written at home can be keyed in quickly when brought to work.

(viii) Provide handbooks on the use of the machine and each program which are simple to understand and contain examples for checking purposes.

(ix) Where possible ensure that the machine becomes part of office practice, i.e. it is identified as part of the work routines.

(x) Link with other practices that have similar machines and are at a similar level of expertise for the exchange and development of programs.

"I must create
a system or be
enslaved by
another man's¹

Wm Blake
Jerusalem

CHAPTER 4

Applications, models and systems

In chapter 2, when dealing with the nature of building appraisal techniques, the concept of a problem-solving 'system' was introduced as a means of dealing with numerical problems (see fig. 2.4). It involved collecting data, structuring the data, using the information in a model or models and finally implementing the solution derived from the model(s). Eventually the solution itself would provide further data which would then be fed back into the system for more structuring, analysing etc. It is an 'iterative' process and by that we mean that the activities within the system are in a loop which is repeated a large number of times albeit with changes in the information that is used.

Many building problem-solving techniques adopt this pattern of events. Existing buildings or processes are analysed in a specified way which then provides the data for the models which are used to predict and discover better solutions. A choice is made and a building design or process is adopted which is monitored to provide more data for the next building of that type, and then the process is repeated. Occasionally old data will need to be removed to avoid incorrect information affecting the models but this will be replenished by the new data from current projects.

This process has largely been undertaken in the past by manual methods. Cost advisers have for example analysed their tender documents and allocated cost items to the major functional elements of the building. The resultant elemental analysis was then used to forecast the new cost of a project using simple measurements of the building (a form of model). Ultimately a satisfactory design was prepared and this was tendered for and then analysed into elements, thus closing the iterative loop. Architects can undertake a similar process when, say, analysing the heat loss of buildings or when analysing spans of roofs and floors to establish suitable dimensions of beams. In each case a hypothetical solution may be tested. If that solution fails then a new hypothesis is tried until a satisfactory insulation value or depth is discovered. Contractors may record data to provide information on, say, the performance of a bricklaying gang which is then used for the next tender and is subsequently monitored to ensure that the model is predicting reliably.

In addition to this fairly linear 'closed loop' approach to

problem-solving, the building team also is able to use several general sources of information which they access and use for their own particular requirements. The common data base may be a library of technical literature which contains information on specification, thermal performance, cost, weathering etc. Models developed by each professional are then used to access the relevant piece of information, analyse it and then use it to assist in the decision-making process. The process may still be iterative but part of the data source may be external to the individual professional's system (fig. 4.1).

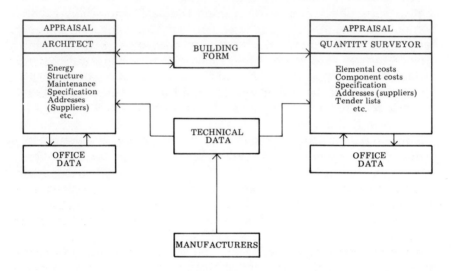

Fig. 4.1 Influence of external data source on iterative processes.

In nearly all systems related to building appraisal there are at least three clearly identifiable parts:

1. The data base.
2. The model, in which the data is used to appraise the building.
3. The monitoring and validation process which checks that the model was behaving correctly and provides useful feedback to the data base.

This applies whether the system is manually operated, computer based or a combination of the two. A recognition of these processes is essential, particularly when it comes to designing software systems for microcomputers. It is of great importance that the data can be easily collected, sorted and classified for use in the predictive models that may be written. The models themselves should be able to be used at the appropriate time in the design process and be able to access or refer to the relevant part of the data base. A method should also be derived for validating the model and retrieving information for future use.

Systems

Before looking at the design of a simple system in detail it will be worth spending a moment or two describing what we actually mean by 'system' and 'the systems approach'.

One reason for the major advances in technology seen over the last couple of centuries is the adoption of 'reductionist' strategies in dealing with human problems. By this is meant the breaking down of complex tasks into simpler, more defined and more specialised tasks. Evidence for this can be seen all around us, in the specialised papers in scientific journals, in the division of labour in factory production lines, and in the building industry where we have seen the growth of specialist professions and trades. The penalty for concentrating on single parts is very often a loss of perspective of the whole. For example if a bricklaying gang was studied and 'optimised' in its working method but the relationship between the gang and other trades ignored, then the results would not be very beneficial and in fact likely to be wrong when applied on site. The interruptions for the plumber, carpenter etc. have a major effect on the performance of the bricklaying team. To ignore these 'buffers' between trades would be to the detriment of the smooth organisation of the building process. Similarly to design a computer system which ignored the relationship between the data base and the models that accessed it would mean that the flow of information would be impaired and that possibly the models would not work at all. The word 'systems' is often taken to mean either:

1. A complex whole, a set of elements or set of connected things which have a relationship between them; or
2. A method or organisation, or considered principles of procedure.

The systems approach

In computing terms what we are generally considering when we talk about software systems is the total package including the information base for the computer models, the models themselves and the processes by which the various pieces of data and the models interact. The 'systems approach' is to take an overview of the whole of the problem to be solved and to make sure that there is sufficient compatability and flexibility in the design of these component parts to ensure the data is handled efficiently and easily to achieve the desired objective of the system. Many writers define 'the systems approach' by attempting to write down a list of stages in tackling a problem. However this is rather a restrictive manner of describing what is intended. The simplest way to describe it is to say that in tackling any problem, the most important part of the work is defining *what the problem is* and then ensuring that the rest of our efforts are used to solve that problem and not some other interesting sub-problem. When in the initial stages of defining the system and its objectives, in planning the strategy, in collecting information, in communicating the conclusions, in getting the conclusions translated into action, we should at every stage be asking 'Is the problem dominating my thinking or is it something else?' In computing terms it is very easy to get sidetracked into placing emphasis on those aspects of the problem which can more easily be calculated and presented rather than dealing with the real end result required. It was argued in the early days that a number of computer-aided design systems had only limited use because they were geared to what is expedient for the machine rather than the user.

It is helpful in designing programs to look at all systems as being in practice subordinate to the large system we call the physical universe. Any system we develop will therefore be a very small part of that larger system. We still have to impose boundaries to our system to avoid undue complexity. We therefore have to be selective in what we include in the system and we have to be aware of what is significant to our problem. We also need to be able to identify in computer systems what degree of computer/human interaction we require, to provide the best end result for our stated objective. There is always the temptation to over-computerise even when we suspect that the human brain can solve the problem faster and more efficiently. For example, despite some of the interesting research work evolved in spatial organisation of buildings, human beings are still faster and more adept at arranging spaces to suit human activity than a machine. It is therefore more appropriate at the present time to let humans organise the arrangement of spaces and allow the machine to undertake the 'number crunching' evaluation of other things such as cost, heat loss, daylighting etc., at which it is faster.

One last point that should be made is that alongside the software system must be the computer hardware upon which it operates. This hardware is also described as a 'system' as again it is a set of parts in interrelation. Some of the factors concerned with the choice of the hardware system have already been discussed in chapter 3 but it must be further emphasised that the software system will often be constrained by the hardware and therefore it is important to have a very clear view of what the total problem is before you start. This will ensure that a machine is purchased which will adequately cope with the software system required. The matching of these two systems for non-standard packages, i.e. programs to be developed in the office, is a skilful task and requires a full understanding of both computing and building work to cope with future possibilities.

Models

Within most systems and in fact in all computer systems there is a need to represent the problem in an appropriate way so that it can be rigorously tested and tried. This is achieved by the use of 'models' which are merely representations of a real life situation. They may be physical models as in the case of an architect's balsa wood model of his building or they may be symbolic as in the case of a mathematical equation or computer program to represent heat loss, acoustic properties or building cost. In some cases it may be the system itself that is being modelled so that the interrelationship between the various parts is tested by some form of simulation. In other cases a variety of models of different types will be brought together to represent various parts of a total system.

In every case the purpose of building the model will be to avoid either loss or expense in building a prototype which can be thoroughly tested. It is seldom practical to build a prototype building to test it for heat loss, cost, daylighting, structure, function and so on before constructing the one that the client will occupy. We therefore use drawings, balsa models, formulae and other methods to represent the building for evaluation. Where the system itself is being represented the model may take the form of a flow chart which is merely a list of operations or activities within the system showing the flow of information and order of operation by means of lines and arrows connecting the various activities. The flow diagram of a network analysis and critical path is an example of this type of model.

There are a number of objectives in using models and these may include:

1. To communicate facts about the system.
2. To communicate ideas about the system.
3. To encourage innovation in the design and operation of the system or building.
4. To predict how the system will behave under different circumstances.
5. To provide insight into why the system behaves as it does.

The word 'system' has been used here in its widest sense and could mean a method or procedure for action or it could mean a physical entity such as a building or component.

The object of the model in a microcomputer system will be to represent the problem by programs and harness the power of the machine to investigate and probe the solutions that exist for the members of the design team. It will generally, although not exclusively, be concerned with item 4 above, i.e. prediction of performance, and because of the current limitations of memory size in the micro it will largely be concerned with numerical analysis. The graphical representation of buildings and the retrieving of large amounts of written specification consume very large quantities of memory store of which the microcomputer is often short. Although this may be overcome by the use of backing store such as disk and overlaying technique, it can result in a rather inefficient process.

Nature of building appraisal systems

To recapitulate: from the point of view of the computer system, building appraisal can be divided into three main stages:

1. The information required to undertake the appraisal, i.e. the data.
2. The models required to represent, measure and evaluate the proposals.
3. The monitoring of the results of decisions taken and based on the models.

This is shown diagramatically in fig. 4.2.

Although this is shown as a linear process it should be recognised that the collection and structuring of data may be going on all the time. Some of the data may be derived from the models or fed back from the monitoring process and therefore there will be a number of iterative loops within a system. In addition, decisions may be taken at various times during the appraisal process with the models operating at various levels of complexity and sophistication as the design is refined.

Applications
There are of course an enormous number of programs that could be used for information, appraisal or monitoring. However to give an indication of the range of applications table 4.1 illustrates some of the subject areas for which programs are available today and which would be expected to be found within the offices of members of a design team. Not included in the table but of growing importance is the use of the microcomputer to act as an interface between the professional and a physical source of information that may be used

in building appraisal. Such systems already developed include the use of a popular micro to access a file of microfiche technical data, standard details, job records etc., without the laborious task of undertaking a manual search. The microfiche information is quickly found and can be immediately projected on to a screen, read and then copied for instant use. Machines have also been used for accessing other physical devices such as temperature gauges, ventilation equipment and even recording the movement of people in rooms for data collection purposes. These can all be considered to be part of a building appraisal software system. However, for the purposes of this book we will concentrate on a simple case study to illustrate some of the underlying principles behind the structuring of a software system for a microcomputer.

An example system

Summarising the procedure introduced above, when developing a system we must look at the *whole problem* and not just the individual parts. We must also be prepared to *set limits* to the problem, which means that we have to decide which factors are *significant* to the solving of the problem. This in turn will mean that we shall *need to be selective* about the data we hold and use. The data we use will largely be *dependent on the models* that we are adopting to predict, monitor, appraise and simulate the problem. The models themselves will be dependent on the appraisal techniques available, the data available, the current state of knowledge and the *limitations of the machine* to store the information required, process at speed and undertake the problem in hand. Not all problems are suitable for machine operation. In this respect we also need to decide *to what degree will there be human interaction with the machine.* Will the machine be allowed to make the majority of choices and follow a standard procedure or will the operator be able to dictate these factors? As you can see it is a complex set of decisions that needs to be made before we can even determine which strategy we are going to follow. Members of the building design team will recognise the majority of these matters as being part and parcel of most design method approaches. The main difference between the manual system and the computer system is that the computer will want to have all these factors determined before starting the problem and it will need fairly rigid rules and procedures to follow in order to obtain a solution.

Defining the problem

Let us have a look at the construction of a simple software system for building appraisal. To illustrate the points to be raised we shall

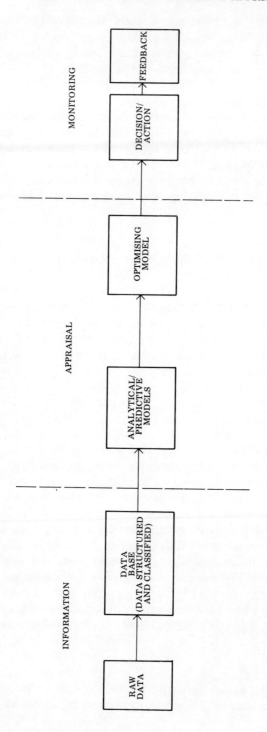

Fig. 4.2 Computer systems.

Table 4.1 Examples of available programs within a building appraisal system

Information	Appraisal	Monitoring
1. *Data file creation/management programs.* These include standard packages such as VISICALC which allow files of information to be set up and manipulated in various ways e.g. totalling of columns, multiplication of rows etc.	1. *Geometric and spatial planning models.* These programs allow the building geometry to be input and held for evaluation.	1. *Office administration.* These programs are not usually part of a building appraisal system but are largely concerned with monitoring office costings and providing standard forms for salaries, valuations etc. Also included here would be word processing packages.
2. *Data structuring programs.* A typical example here is the breakdown of a bill of quantities into various formats for elemental analysis, NEDO formulae and stage payments etc.	2. *Building performance models.* These programs evaluate the geometry or some other aspect of the building. Included under this heading would be heat loss calculations, daylighting, cost predictions, cost-in-use, structural calculations etc.	2. *Reconciliation programs.* These take feedback from site, office etc., and compare with the planning and performance models. Included here are valuations related to S-curve predictions, project planning, price adjustment formula etc.
3. *Data search programs.* These programs allow rapid access to relevant files or items of information.	3. *Planning models.* These programs assist in the organisation and prediction of procedural and financial matters and include such programs as critical path analysis, S-curve predictions and development appraisal.	3. *Analytical monitoring programs.* These programs analyse the feedback on performance to discover why events have occurred. They may also be attached to programs which will assist in correcting any problems found in monitoring the building process, e.g. reworking of a cash flow or network.
4. *Data sort programs.* These programs sort the data into a specified order.	4. *Statistical models.* These programs analyse data to detect trends and describe the nature of the data. They include regression models, measures of central tendency, dispersion etc.	
	5. *Simulation models.* These programs simulate changing events over time usually by sampling from specified distributions.	
	6. *Optimisation models.* These programs attempt to find the 'best' solution according to specified criteria.	

be referring to the example of the building envelope although the principles can be applied to many aspects of building appraisal. Even though we shall be considering this element of the building in isolation we should note that the design of the envelope will have a considerable effect on other components within the building including partitions, lighting and heating components, service runs, structure and spatial organisation. Conversely the envelope itself may be dependent on these other factors. This interdependency between elements of the building system is very difficult to cover within a small microcomputer system because of the extensive amount of cross-referencing required which makes large inroads into the available memory store. However in a total building appraisal system an attempt should be made to consider automatically the repercussions of a decision through the total system or, at the very least, to place warning notices in the computer output. For our purpose we will assume that something of this nature will be provided. The following notes are not intended to be a comprehensive guide to the development of a system but rather an introduction to the concepts involved.

The data requirements

If we go back to our requirements for the system we said we needed the data, the model and possibly some form of monitoring. There is of course a great deal of interdependency between the data and the model — it is not much good building a magnificent model if the data is not available to use it, and conversely it is not much good providing an extensive data base which is not in the right form for the model. In the case of the building envelope we know that we probably have reasonable data available on such matters as:

1. U-values for glazing, claddings and roof.
2. Unit costs for major components.
3. Weights of components.
4. Maintenance characteristics.
5. Wind resistance.
6. Structural performance of materials.
7. Ranges of finishes available for claddings.
8. Insulation values.
9. Operating costs (e.g. window cleaning).
10. Building Regulations.

In a large system these items would be a series of major files which would be common to all or a large number of building projects and which can be accessed when the respective information is required.

This would be part of the general data base. In a microcomputer the limitations of memory size and the slow speed of access of information which we must accept at present make this kind of large data base impractical unless the machine can be harnessed to a mini-computer with large backing store. We therefore need to reduce the amount of common data held and to make it more specific. This can be done by limiting the data to a single building type, or omitting common data altogether and inputting data related only to the job in hand. This however can be a very time-consuming task as every piece of information to be used in an appraisal will have to be keyed in for each new project and it should be observed that the machine can often lose its credibility with professional staff unless the computational aspects of the system are seen to be dramatically advantageous.

In addition to the general information a unique data base will be required relating to the project in hand. In this very specific set of files will be held the further information that is required by the models to appraise the building, or in our case the envelope. These files could include items such as:

1. The building shape and form (i.e. its geometry).
2. A description of the elevations with the amount and type of glazing and the various claddings.
3. A description of the roof structure, covering and form.
4. A description of the building orientation and exposure.
5. Pertinent climatic conditions relating to the site.
6. Quantitative measures of the major components (areas, number etc.) This may be generated by one of the models depending on the way in which the building shape is input. If co-ordinates and dimensions are given then measurements can easily be generated.
7. Client attributes. This may include details of the client's tax position for the purpose of discounting future costs or revenues.
8. Site and other constraints which may well affect the shape of the envelope and also have repercussions on Building Regulations affecting the position of the building.

Once again we have to bear in mind that we are usually short of main memory and even backing store will be limited. Therefore it may be necessary to arrange the appraisal model (assuming all this specific data is called for) in such a way that only relevant data is accessed when needed and then removed or returned to the disk file to allow new data to be input. It may even be essential to have data on more than one disk which is changed by hand when required in the appraisal program.

For the purposes of this simple example system we shall limit ourselves to just three appraisal programs, one for heat loss through the envelope, one for cost and one for cost-in-use.

The data base

Having looked at our data requirements we may well have a data base that looks something like fig. 4.3. These are the files that will be accessed by the appraisal programs to provide the information to produce an evaluation.

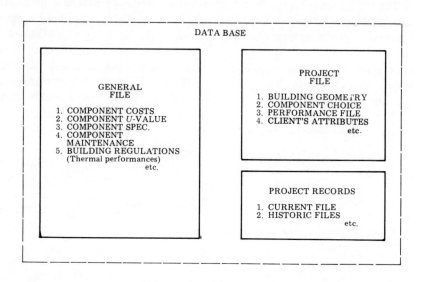

Fig. 4.3 Data base files.

The general files will be held and updated on disk and will be available for any program that wishes to access them. For simple general data of this nature it is common to 'string' the data together in a predetermined manner, rather like a table of component properties. For example it may look something like this for each component:

1	2	3	4	5	6	7
32 mm single glazing	5.00	5.50	20	6.50	1.00	—

◄—field width——►
(e.g. 20 characters)

—►field ◄—
width
(e.g. 4 characters)

Column 1 = description
Column 2 = U-value
Column 3 = cost per square metre
Column 4 = expected replacement period (years)
Column 5 = expected replacement cost per square metre (at present prices)
Column 6 = operating costs per square metre (e.g. cleaning)
Column 7 = weight of component (not used here).

Immediately it can be seen that in compiling such a table some problems have arisen. To describe and identify the glazing properly we would need to enhance the specification which would take up a considerable amount of memory space. The other properties have been included as fixed amounts but these factors in real life may vary quite considerably. Costs are not static and will vary according to size of pane. Assumptions with regard to life of components or even U-values are seldom correct. Ideally, distributions of cost and performance should be included but once again we are likely to come up against limitations of memory store. In practice we would have to weigh up the advantages derived from more information against the extra load on the memory. In making this decision it would be necessary to consider the design stage at which the information would be used. If it was intended to use the information all the way through to detailed design then we would approach the problem differently than if we recognised that it would be restricted to feasibility studies.

However, the advantage of such a table is that providing all components are structured in the same way, i.e. the width of 'field' of each column space is identical, then a universal program can be written. This program would be able to go to column 4 in each case and find the expected replacement period no matter what component it was considering; it could then use this piece of data in its calculation of cost-in-use or whatever evaluation it wishes to undertake.

To supplement the general data files would be a series of programs which would allow the files to be revised, removed or sorted in a particular way. Thus every data base will have a 'system' of its own to maintain its usefulness and keep it up-to-date.

The project files would tend to follow the same pattern but of course by their very nature are likely to contain different data for each project. Consequently these can be set up as far as possible at the beginning of each job or items of data can be added to them as the programs are run and information is input via the keyboard. In practice it is not unusual for a history of the project development to be required and therefore files recording the decision-making process and the results may also need to be kept. A number of possible building geometries may be held, together with the results of their evaluations. This requirement would demand that all proposals were date-marked in order that progress could be monitored and possible back-tracking through the data could be achieved.

The structuring of data bases and the speed at which information can be input, accessed and retrieved are vital to an efficient system. For this reason a simple introduction has been included in chapter 5 alongside the manner in which the problem can be logically represented. Chapter 9 gives a practical guide to the programming required for a simple data base.

The programs

The importance of the data base has been stressed but there are three other elements which need careful attention and these are the operating or master control suite of programs, the appraisal packages and the monitoring programs.

The software master control system

The operating programs will play a major part in making the total system 'user friendly' or not. From the time you switch on the computer and call up the system part of its job is to lead you through the choices of programs available — it should explain how the system works and prompt a response in the required manner. The gauge of its success will be found in the efficiency with which it accesses the programs, the lack of ambiguity in the choices and instructions available and the help that it provides for the novice user.

In the case study that we are considering the main concern of the master controls will be to provide:

1. Access to the data base routines for updating and checking.
2. Access to the appraisal packages.
3. Explanatory messages regarding the operation of the system.
4. Identification of errors in the input.

It may also handle the interrelationship between the programs within the system.

An overview of our proposed system can now be shown diagrammatically in fig. 4.4. In practice as far as the operator is concerned he will probably only be aware of the operating system by the sets of choices of program (sometimes known as menus) presented to him.

In our example it is likely that the first menu will include these sets of options:

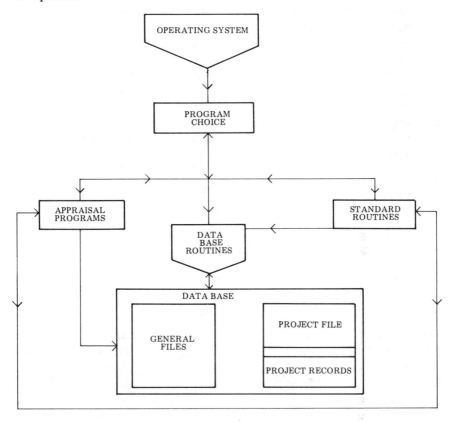

Fig. 4.4 Overview of building appraisal system.

MENU 1

MAIN MENU

PLEASE CHOOSE FROM THE FOLLOWING:

1. Change, delete, add to the general data base.
2. Set up new project files.
3. Change, delete, add to existing project files.
4. Review existing project records.
5. Run appraisal packages.

If selection 5 is chosen by entering the figure 5 from the keyboard then a new menu will appear giving another series of choices:

MENU 2

> PLEASE CHOOSE FROM THE FOLLOWING FOR
> APPRAISAL OF THE BUILDING ENVELOPE:
> 1. Energy appraisal.
> 2. Initial cost.
> 3. Cost-in-use.
> 4. Return to main menu.

Item 4 is included in the above display to allow the user to return to the original options and subsequently follow a different route through the system.

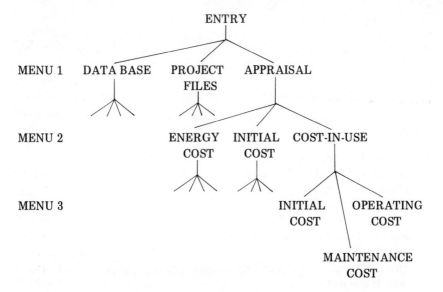

Fig. 4.5 Hierarchy of operating system, showing the result of following one particular path.

When one of these choices is made the menu for the particular appraisal program will appear. By structuring the choice in this way the user is led through the system with all his options made explicit. This technique is 'hierarchical' in its format, i.e. as each selection is made it gradually narrows down the range of choices until the user's intended program is found (see fig. 4.5). Once found, the machine loads the relevant program, and possibly its data from disk, and commences operation. In this way we can avoid holding all the programs at one time and avoid overloading the memory.

The appraisal packages

We are now at the point in our simple system where we can consider the evaluation of the building envelope. This will in practice involve us in the construction of what we would hope to be 'robust' models. By robust we mean a model that will cope with all the demands we shall make upon it within the limits that we have defined for our problem and the available data, and constructed in such a way that checks can be made against these constraints to ensure they are not exceeded. In addition all commercial computer programs should include extensive checks to give the opportunity of correcting erroneous input from the user before the program advances.

THE BUILDING GEOMETRY

So far we have talked about data in rather general terms. When we come to construct a model it is necessary to be very specific about the information we need to use and which therefore needs to be found in the data base. For example, whereas we have referred to building geometry, we now have to decide exactly how we are going to hold the geometry in the memory of the machine in order that we could use it for appraisal. There are a number of ways of tackling this — we could:

1. Draw the plans and elevations on the screen, 'dump' (i.e. transfer and record) this image to disk and use a sub-program to measure areas, perimeters and volumes from this image. A number of advantages would accrue from an architectural viewpoint but the method may require very complex programming, e.g. how does the machine recognise a boundary or void on the screen? It will also require a large memory because it will hold the whole of the screen, even having to record the blank spaces which are not used.

2. Store the plans and elevations in modular form within a matrix in the memory store where every 'box' in the machine represents one unit of our building. Where the 'box' contains a line it represents a boundary of unit length and the empty 'boxes' represent the blank areas between (fig. 4.6). We can calculate areas and perimeters merely by adding up the boxes. This method is similar to the previous suggestion and has the same drawbacks but with the likelihood of an increased loss of accuracy due to the modular nature of the boxes, i.e. it is difficult to specify half a box.

3. Use co-ordinates to describe the building. If the position of corners, angles and so forth are defined by their co-ordinates in relation to some reference point then measurements can be

	A	B	C	D	E	F	G	H
1								
2		X	X	X	X	X	X	X
3		X						X
4		X						X
5		X						X
6		X						X
7		X		X	X	X	X	X
8		X		X				
9		X	X	X				
10								

Fig. 4.6 Plan geometry held as physical elements within a matrix. X represents the outline of the building.

derived by normal geometrical principles. It is of course a time consuming business discovering and inputting all the co-ordinates, but once held this method does enable rapid alterations to the geometry to be made by overwriting, adding or deleting previous co-ordinates. In terms of memory space, it is much more efficient than the two previous systems but this needs to be weighed against the extra time for input.

4. Merely enter the relevant area, lengths and volumes required by our model. This obviously means that a fair proportion of the measurement and calculation work undertaken automatically with the three previous methods will now have to be done manually. Any changes in geometry will also require some manual calculation before the change can be incorporated in the model. It is fair to say that it may be far quicker for the user to take off some coarse measurements where little anticipated change in the geometry is likely to occur than to write and run a program to do it. Human measurement will avoid identifying co-ordinates or inputting drawn information and will save on memory space and length of program. Unless the geometric information is to be accessed for a wide range of uses and in particular plotting of the results, then it is unlikely that the more laborious type of input will be economic.

It will be appreciated that in an appraisal of the building envelope, the choice of geometric information will be critical to the whole system and will to some extent influence the whole of the appraisal package.

Similar decisions, but probably not as complex or far-reaching, will have to be made about the other inputs to the project data base. The important things to bear in mind are:

1. How much data do I need to input in order to get a satisfactory result from my appraisal packages? The balance here is between the extremes of sophisticated data that may never be fully used and coarse data that may produce unreliable results.
2. Am I likely to change extensively and manipulate the data as the system operates? If so, I can trade off the initial time consuming input against future time savings when making alterations.
3. Which system is most likely to suit the kind of operator and operation which I envisage within my firm? If complex input is likely to discourage staff then it may sway the decision towards a more simplified approach.

THE APPRAISAL MODELS

In some respects the appraisal models are relatively straightforward once the data requirements are sorted out. Since the data is in some cases dictated by the models and in others the reverse applies, a chicken and egg situation appears to develop, but in reality the two are usually considered together and common sense provides a quick answer. It is normally the data which is first to be structured in detail in order that the model can be written to access it.

In the first instance it is likely that the system designer will look at the manner in which the manual model operates for building appraisal. Hence if we are considering heat loss through the external wall, then a simple formula may be the starting-point, such as:

$$\text{Heat loss} = \text{area of wall} \times \frac{U\text{-value}}{\text{for wall}} \times \frac{\text{temperature}}{\text{difference}}$$

To this calculation might be added other factors such as degree days or equivalent hours to give annual heat loss which could then be translated to its fuel equivalent and priced and capitalised for use in a cost-in-use model. This might be the traditional manual method for a simple solution and it is easy to write such formulae into a program to save on manual calculation. It may be that additional refinements can then be added by, for example, accessing the data base and allowing the operator to choose from a selection of wall types which also have their attributes attached to the description.

As you pick a wall type, you also take with that description a cost and a *U*-value together with other pieces of information which are then put into the formula without any manual operation. The measurements for the wall can be taken from the building geometry in the project data base and consequently with the user actually making only one choice a comprehensive evaluation routine can be operated. The computer can now be seen to be coming into its own but we have still not realised its full potential. We know that machines are tireless calculators and the next step could be to take advantage of this capability to try out any number of options. If the computer accesses each wall type in the data base in turn and undertakes the calculation then a table of results can be produced which will compare the heat loss and fuel cost implications for all the wall options. The designer can then make a choice from the information provided knowing the energy and cost implications.

A simplified chart of these operations can be seen in fig. 4.7. In practice this chart would need to be amplified to include checks against other criteria such as the Building Regulations and also against incorrect input by the user.

SUB-ROUTINES

As the model becomes more sophisticated it is likely to use a number of different routines within the appraisal. Many of these routines will be standard to a large number of building problems e.g. discounting, sorting, outputting screen contents to disk file, data input checks and so forth. It would therefore be sensible to have these standard routines available for use in any program. At the appropriate time in the appraisal the required sub-routine would be called up to undertake its task and then be returned to disk. If space is available it may be possible for the routine to be permanently attached to the program thus saving time reading and writing to disk.

Most programmers build up a library of these basic program 'building blocks' which they know from experience will be used time and time again. They are usually jealously guarded and regularly refined rather like a good tradesman would look after his tools. Their aim will be to make them as reliable and fast as possible as well as to make them of universal application; the programmer's ability in this respect is a good measure of his or her aptitude and ability as a programmer.

Monitoring and validation

All models constructed on the computer need validating at some time. Reasons for this include:

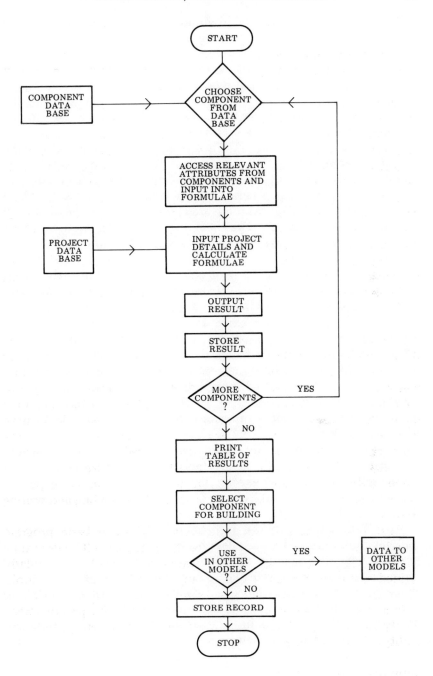

Fig. 4.7 Simplified general model of appraisal process for selection of components.

1. Checking for errors in the program.
2. Checking for errors in human input.
3. Checking for data base errors.
4. Checking for reliability of the model.
5. Checking for usefulness of output.

Very few programs are written without someone saying, 'Well I would rather like to add this option . . .', or, 'Couldn't we improve the performance if we . . . ?' We are at the embryonic stage with regard to computer modelling, at least for building appraisal, and this inevitably means that considerable improvement can take place.

One reason that the programs at the end of this book have been called 'starter packs' is that there are any number of ways of improving them. The reader can take them and develop them in whichever ways he likes to suit his own ability and the limitations of his own machine.

Getting the program right is of course fundamental to the monitoring and validation process. However monitoring can also mean reconciling real life situations with either the model or its data base. Perhaps the most simple example of this is comparing monthly valuations on a building project with a projected cash flow for which the operator needs to access the project records to find the forecast and then to plot the real valuations against the expected figures. From the plotted comparison other techniques can be used to provide a current view of the expected completion date. A simplified version of a projected cash flow program is included in starter pack 1.

The monitoring does not of course have to be financial. If temperature gauges are fixed to the fabric of a building and recorded readings fed via an analogue to digital converter to the micro it is possible to monitor heat flow and compare this with the forecast of the model.

In nearly every case the historic information contained in the project files will need to be accessed. As most appraisal models will already have the facility to call up past records for revision purposes it will often be sensible to use that part of the program in the monitoring routine; indeed, a good approach would be to have an option for the monitoring of models in the first menu that is displayed when commencing the operation of the system. Calling records from file would then be a sub-routine which the monitoring program would use alongside its own input, output and analysis routines.

Testing models can be extremely time consuming and monotonous, but it can be a very skilful task involving clever detective work to discover if and where things are going wrong. In addition a good deal of ingenuity is required to devise suitable methods of validation as it is often difficult to produce a satisfactory test when measuring an item involving human performance. This performance is usually so variable and dependent on so many factors that to compare just a few real life situations with the model will not necessarily yield satisfactory results. Statistical methods are available for these more difficult tests and could form part of the software system but they are beyond the scope of this book.

In respect of our building envelope system, monitoring is likely to be limited to comparing:

1. Predicted cost with actual cost.
2. Predicted heat loss with actual heat loss.
3. Predicted component performance with actual performance.

Although item 3 has been included, in reality it is usually an impractical proposition as replacements and repair extend far into the future, by which time most clients will have forgotten about the original prediction.

Any feedback from the monitoring should in theory be used to fine-tune the appraisal models and/or the data base. Whilst this can be achieved fairly easily in certain areas, e.g. revisions to unit costs, it is much more difficult in other appraisals where the information returning from site is less firm and in many cases is vague and confused. This would apply to heat loss where it may be very difficult reliably to determine the heat loss through say the external walls.

The final view of our system will therefore tend to look something like fig. 4.8. The diagram shows the interrelationship between the major elements of a simple system for what is a fairly straightforward appraisal. As the models become more sophisticated requiring more refined data and sub-routines, so the system would grow and become more complicated. However the basic principles and concept outlined above remain valid. The complexity will come from the type of models used and, more likely, the interrelationship between the models themselves and between the models and the data base.

Summary

The general principles involved in developing a software system can be summarised as follows:

1. Elements within the system should be accessible to other related elements of the system. (This is what makes it a 'system' rather

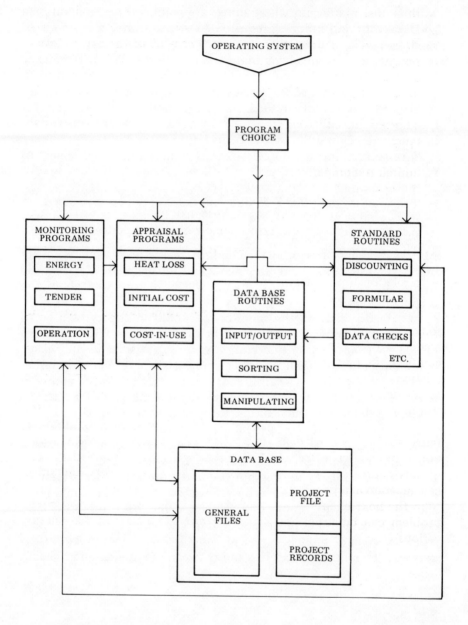

Fig. 4.8 Simplified software system for building envelope appraisal.

than a series of *ad hoc* programs.)

2. Data within the data base should be properly structured and addressed in a uniform manner for ease of access by any other program.
3. The operation of the system should be such that the user can quickly and efficiently access and operate any part that he chooses.
4. The system should be able to be expanded when required to encompass new appraisals or data.
5. The system should suit the level of information available for appraisal and should therefore be related to some form of design method which can define the levels available.
6. The models and data should be capable of being refined as more information becomes available and techniques are improved.
7. The interrelationships assumed between elements of the system should be made explicit, i.e. the programs should identify where a relationship exists.

In this chapter an attempt has been made to introduce the reader to some simple concepts related to software systems. In practice the reader will find that the development of such a system can be very complex. In particular the recognition and structuring of the interrelationship between elements can result in stretching human logic to its full. Having said that, it is unusual for large complex systems to be developed on microcomputers; the normal pattern is for one or two appraisal routines to stand in isolation or to access a fairly straightforward data base comprising a simple table of components together with their attributes. Even so, a simple data base of this nature can provide a powerful source of information which can generate a large number of evaluations.

In the next chapter we look in more detail at the way an appraisal problem can be structured and the effect of structuring file data in various ways.

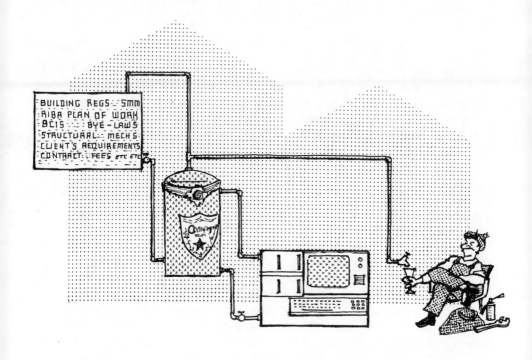

BUILDING REGS   5mm
RIBA PLAN OF WORK
BCIS   BYE-LAWS
STRUCTURAL   MECH &
CLIENT'S REQUIREMENTS
CONTRACT   FEES etc etc

CHAPTER 5

Problem structure and file organisation

In the previous chapter we introduced the concept of 'the system', an important aspect of any computer work. Without an understanding of how various programs dealing with a common focus for evaluation, e.g. a building, can be interlinked and interrelated it is difficult to achieve the full potential of the microcomputer. The key to this potential lies in *the efficiency* with which the memory and operating systems are used. Considerable care should therefore be given to the design of the system and its expected future development.

It was noted that in building appraisal, as in so many other computer applications, it is the data base which aids and assists in this building up of the interrelationships. Its structure, content and input/output routines play an important part in satisfying the operator of an appraisal system. Later in this chapter we shall look at some simple ways to structure data but first of all we shall look at how we approach any type of problem solving on the computer.

Computer problem-solving

We all recognise that in building appraisal we are concerned with the solving of a problem. 'How much energy demand does this building call for?', or, 'What will this shape of building cost?' These are questions which derive from a demand to know how the building will perform in practice. When a client commissions a building he poses a problem to his design team which they need to solve before the building is built. Hence the need for models to represent aspects of that problem which can be tested. As we saw in chapter 2 there is a difference between the way in which humans and computers solve problems. One important consideration is that the computer will need to know in unambiguous terms exactly how to tackle the problem. It will need to know precisely what data is to be input, what calculations need to be done and in what order. On top of this it will need to know the questions that need to be asked, the layout of the questions on the screen and exactly how the final results are to be presented.

Humans undertaking the same task will allow themselves much greater flexibility, asking questions when it suits them, making judgements relating to the importance of information and writing out results in an individual manner. In view of the rather rigid logic of the machine it is necessary to define the detailed order of events in the problem-solving process before starting to write the program. Even in manual problems it is often a useful exercise to rigorously define a standard procedure for people to follow. The discipline is useful because it avoids things being forgotten, aids checking and assists in comprehending complex procedures and output.

Approaches to programming

Most programs that relate to building appraisal are likely to have three major parts:

1. *Input data.* This is the information necessary for running the program and it may be derived from any of the input devices, e.g. keyboard, disk backing store, digitiser etc.
2. *Processing.* This is the manner in which the input data and any standard data held within the program is manipulated in order to achieve a solution to the problem.
3. *Output information.* After all the hard work has been done by the operator and the machine this is the information that is required to be presented from the end result of the processing. It can be output to a number of different devices including the VDU, the printer, plotter and in some cases to the backing store where it will be used by other programs.

The program may not necessarily follow the above order in its execution, except for very simple problems. Usually there are a number of interim outputs and inputs, with data being called up from backing store or via the keyboard in order that the next part of the program can be processed. Indeed some of the keyed-in information may be based upon the results of the previous piece of program that has been executed. As the problem becomes more complex, so it becomes essential to think it through right from beginning to end.

Procedure for programming

The following general approach may assist in this task:

1. Define the problem.
2. Where necessary provide a mathematical expression for the problem.
3. Determine the kind of output required.

4. Determine the inputs required to achieve the above output.
5. Design the program logic to produce the desired results and output.
6. Structure the program in sections.
7. Write the program.
8. Debug the program.

In addition it will be necessary in many programs to actually validate the model used (particularly if it differs from manual technique) and to test its implementation into the office routines. Too many programs have failed, not because they are poor at calculation or the model is inadequate, but because they are not suited to office practice. There is a psychological barrier between many people and the machine, and unless this is borne in mind the most sophisticated problem-solving routine will never get used. Assimilation time for the operation of a program is important to its credibility. The building industry is littered with programs that people tried to use, found difficult, and therefore rejected. This factor is very often underplayed and should be brought to the fore at each stage of the general approach.

Defining the problem
In its simplest form the problem may already be well defined (see the example given in chapter 6). The various small parts of building appraisal, such as structural calculations, energy demand and cost prediction, have all been undertaken manually for a considerable number of years. The difficulty arises when a new idea is put forward for tackling the problem based on another premise. For example it has been recognised for a number of years that energy demand is not just the direct result of fixed assumptions with regard to air changes and U-values. The behaviour of materials and more importantly the behaviour of people within the building has a considerable impact on the final energy demand.

If this variability of human behaviour is to be taken into account then a new model involving the simulation of human behaviour will be required. The problem will have to be redefined in a different way relating to behaviour patterns. With this new concept will come some other questions, for example:

(i) Does the firm think it worth-while to pursue this new model?
(ii) If so, can a feasibility study be undertaken to test whether it is likely to be profitable to invest in its development?
(iii) What form should the feasibility study take? Are there suitable criteria available for testing the new proposal?
(iv) If the feasibility study proves the idea to be acceptable, can

the problem be defined sufficiently to brief a systems analyst and programmer?

(v) Is there likely to be sufficient data available to run the new model?

This last point is vital in problem definition. Many of the research projects undertaken in universities and polytechnics in the UK over the years have produced sophisticated models which on paper show a marked improvement on existing practice. These include regression analysis and simulation models for prediction. Unfortunately there has not been enough data that is easily and continuously accessible which will feed the models to keep them up to date. Perhaps if a UK national data bank were to be created such as that envisaged by the Building Cost Information Service then things may be different but there are all sorts of reasons why this is difficult to achieve.

Problem definition is not therefore necessarily a simple statement of objective but it entails a thorough look at every aspect surrounding the problem.

A mathematical expression for the problem

We have already discussed the nature of computer problem-solving and the manner in which it uses a binary number system to handle information. It therefore responds more readily to numerical information and mathematical expressions than to words and speech. Where possible our problems should attempt to use numbers and letters in algebraic form to increase the efficiency of the machine. To take a common example, for microcomputers it is preferable to handle the UK National Building Specification by placing an alphanumeric code against each description rather than use words in the search process. Similarly with a standard library of descriptions for bills of quantities it is preferable to use a faceted code to bring together the various phrases which combine to make a full description in a tender document.

These codes represent a mathematical description of the problem which can be easily handled by the machine. Even in a more straightforward problem, such as the taking of measurements, it is important to consider the measurements in algebraic form. Each measurement will be stored in a unique address within the computer. To undertake a calculation the programmer manipulates the contents of the addresses.

For example

$$A = B + (C - D)$$

means

Take the contents of address D away from the contents of address C
Add the result to the contents of B
Place the final result in address A.

It is not strictly algebraic as there can only be one address on the left hand side of the expression. All calculations undertaken by the machine will take this format and the problem will need to be structured in this manner. If it is necessary to find the area of solid external wall in a rectangular building the expression may look like:

$$A = [2 (L + B) * H] * [(100 - G)/100]$$

where L is the length of the building, B is its breadth and H its height, G is the percentage glazed, A is the area of solid external wall, * is a multiplication sign and / is a division sign.

Whereas by manual technique we may physically measure all elevations, in the computer we need to express the *process* we go through in this mathematical way. This also applies to the logic the computer employs. If for example the computer is required to compare the external wall U-value (contents of address A) with a known insulation standard (contents of address K) and then execute a different piece of program when the standard has been achieved, then the mathematical expression appearing in the program would be simply:

IF A < K THEN GO TO 121

where < means less than and IF ... THEN GO TO is computer language. The expression means that if the contents of address A are less than the contents of K then branch at this point and go to line 121.

The whole of the program will be structured in a similar way. Even words will be held in addresses, identified by alphanumeric characters which will hold what is known in the BASIC computer language as a 'string' of information. Strings are usually identified by a $ sign after the variable (address) name. For example,

```
10   A$ = "MICROCOMPUTER COURSE"
20   PRINT A$.
```

This small program would place MICROCOMPUTER COURSE into an address called A$ and in line 20 it would print out the contents of A$.

These facets of programming are dealt with in much more detail in section II. They are introduced here to demonstrate the mathematical structure of the computer solution to the problem.

Determine the kind of output required

The reader will appreciate that to know where you are going is absolutely vital when structuring a solution to the problem. The objective of the program is not found within the processes it goes through, but in the end result which is finally expressed on one or more of the output devices. To this end it is worth while giving very careful consideration to what is required as the end result. Do you require tables of information; if so, what information do you want presented in that table? Do you require all selections made by the operator to be printed out, and if so what choices will he be allowed to make? These are just two examples of the kind of questioning generated by considering the output.

The authors' approach to computing has always been to try and define an end result before starting any of the other activities. In programs dealing with a single form of appraisal this is relatively straightforward. It becomes much more difficult when dealing with a large system. In large systems there are several programs, data handling routines and utilities which all have their own individual objective but in addition also contribute to an overall objective for the whole system. However the same principles apply and two sets of output can be analysed related to, firstly, each individual program, and secondly, to the system itself.

One aspect of the output that may be required is flexibility, i.e. a choice as to how the information should be presented. For example a program investigating building shape may require outputs relating to heat loss, cost, quantities, cost-in-use, wind loading and structure. It may also be envisaged that at a later date an additional routine may be included in the system and this would need to be accommodated. By giving careful consideration to these points at this stage the future structuring of the program will be more closely tailored to the objective.

It pays to actually draw out draft outputs with all essential information so that significant factors related to the program can be identified.

Determine the inputs required

Having established where we are aiming it is important to identify what information we need in order to get there. To a large extent this will depend upon the models we choose to achieve our objective. What items of data do they rely upon? When do they require each piece of information? What alternatives are to be offered? These questions naturally arise when considering the problem and they will be closely related to the mathematical structure (in essence the model), probably already chosen. The structure dictates that certain information is available at certain times in the program.

It is often helpful to list all the variables that require data to be input and then decide:

(a) where the data is to come from — is it to be obtained from standard data, keyboard, disk-based data base, computed etc.?
(b) when that data will be provided in the program — is all information to be input at the start, for example?
(c) whether the data needs to be held for record purposes.

It is surprising how often in building appraisal that records need to be kept or assumptions made explicit so that other members of the design team know the basis for a particular result. The computer can rapidly provide these records and it would be a pity if this additional power were not used to benefit the communication between the participants.

In considering the data inputs you will also be testing whether a particular approach to problem-solving is feasible. We have already mentioned this under 'defining the problem' but its importance cannot be stressed too much.

Designing the program logic
Once the inputs and outputs have been ascertained and the problem described in a mathematical expression we are left with the organisation of these facets into a logical order that will achieve the desired result. By desired result we can mean several things including:

1. The calculations take place in the right order;
2. The data inputs occur at the right place;
3. The screen information is provided at the right time and right format for the user;
4. Records are kept of the interim and final results when required;
5. The right sub-routine is chosen when a particular choice is made by the operator;
6. The right number of iterations or loops occur in the program;
7. The printing of titles and results occur at the right time; and so on.

The manner in which the program logic is normally shown is by means of the flow chart. This is a sort of strategic plan for the programmer and is now familiar to most of us through its use in a number of fields, including newspapers and television, to demonstrate procedure.

There is available a series of standard conventions under British Standard BS 4058 which enables flow charts to be universally understood. A list of the more common of these symbols is given in fig. 5.1.

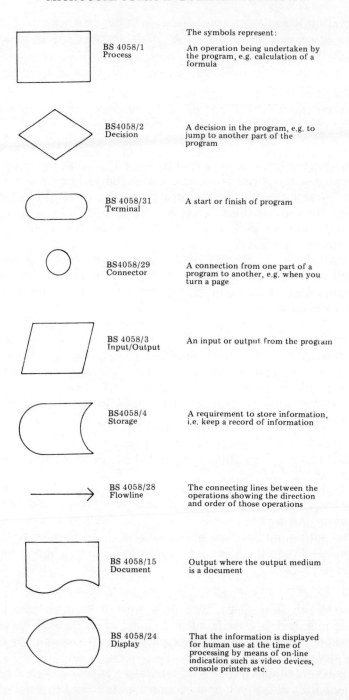

The symbols represent:

BS 4058/1
Process — An operation being undertaken by the program, e.g. calculation of a formula

BS4058/2
Decision — A decision in the program, e.g. to jump to another part of the program

BS 4058/31
Terminal — A start or finish of program

BS4058/29
Connector — A connection from one part of a program to another, e.g. when you turn a page

BS 4058/3
Input/Output — An input or output from the program

BS4058/4
Storage — A requirement to store information, i.e. keep a record of information

BS 4058/28
Flowline — The connecting lines between the operations showing the direction and order of those operations

BS 4058/15
Document — Output where the output medium is a document

BS 4058/24
Display — That the information is displayed for human use at the time of processing by means of on-line indication such as video devices, console printers etc.

Fig. 5.1 Flow chart symbols (from British Standard BS 4058).

The level to which the flow chart is drawn will depend on the systems analyst or programmer. In some cases with simple programs, a flow chart may be felt to be unnecessary. In other cases the problem may be so complex that several flow charts at different levels of complexity may be required. The first diagram may just outline the interrelationship between the programs and the later diagrams will spell out the operation for each program individually. There are no hard or fast rules and experience will eventually dictate what level of detail is necessary for ease of programming. It is important to realise that the chart is a vehicle for communication and provided those who need the flow diagram understand it, and know what they have to do then that is sufficient.

To illustrate the use of these charts three examples have been chosen.

EXAMPLE 1

Fig. 5.2 demonstrates the use of the technique for the very simple operation of crossing the road in the UK, where drivers drive on the left hand side.

The decisions and operations are clearly identified and if the procedure is adopted then the chances of an unfortunate accident are reduced.

EXAMPLE 2

This example shows the procedure that the computer will have to adopt to sort a series of 10 numbers into ascending numerical order. Sorting is a common routine in many programs. Please refer to fig. 5.3.

The operations are as follows:

Operation 1

This reads the numbers input via the keyboard or other device into an array. An array is a row of addresses rather like a row of pigeon-holes with a number placed in each address. In this case we have called the array X.

	1	2	3	4	5	6	7	8	9	10
X	2	1	4	6	3	8	9	12	5	7

An address is identified by its alphanumeric code so $X(3)$ contains the number 4 in the above example. In the computer the numerical part of the code can be a variable so $X(N)$ can be any one of the addresses depending on the value of N. Likewise $X(N+1)$ will be the next address along from $X(N)$.

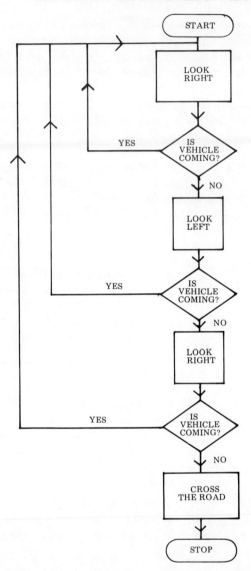

Fig. 5.2 Flow chart for crossing the road where traffic drives on the left hand side.

Operation 2
This sets the value of N to 1 and K to zero so that the sort routine will work (see operations 3 to 5).
Operation 3
This is the part that does the actual sorting. The approach taken is to check whether the number in the first box X(1) is greater than

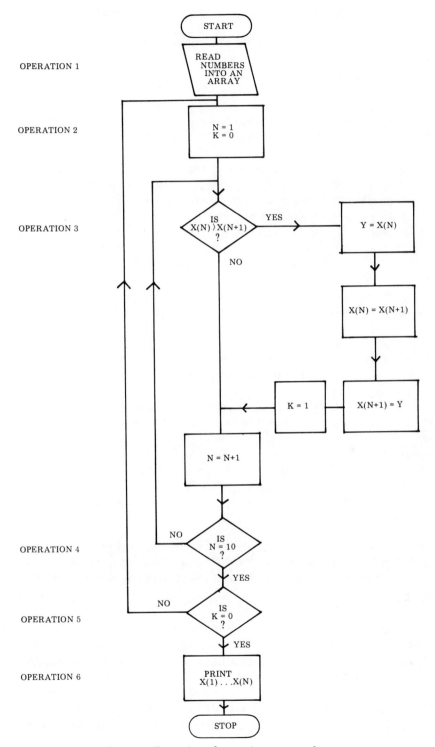

OPERATION 1

OPERATION 2

OPERATION 3

OPERATION 4

OPERATION 5

OPERATION 6

START

READ
NUMBERS
INTO AN
ARRAY

N = 1
K = 0

IS
X(N) > X(N+1)
?

YES

Y = X(N)

NO

X(N) = X(N+1)

X(N+1) = Y

K = 1

N = N+1

IS
N = 10
?

NO

YES

IS
K = 0
?

NO

YES

PRINT
X(1)...X(N)

STOP

Fig. 5.3 Flow chart for sorting ten numbers.

the number in the second box X(2). If it is then the two numbers are changed over in the array, so that these two numbers are now in ascending order. This is done by setting N to 1 and comparing X(N) with X(N+1).

If X(N) is not greater than X(N+1) then the machine increases the value of N by one and leaves the numbers where they are. Therefore next time round the computer compares X(2) with X(3) and so on until all the numbers have received an initial sort.

You will notice that K is set to 1 if the numbers are swopped. This is a 'trip' for the computer to realise that it needs to go back to the beginning to ensure that the swop has not affected numbers lower down the scale (see operation 5).

The variable Y is just a temporary address in which a number is placed to avoid it being lost in the swopping process. Y is overwritten on each iteration by the new value that is placed in the store.

Operation 4

This just checks that all the numbers have been sorted. If not, it returns to operation 3 and continues to compare each box with the one above it in the array.

Operation 5

After all ten boxes have been sorted once and at least one alteration or swop found to be necessary then it returns to the beginning of operation 2 and starts all over again.

It keeps on doing this until no further swops take place; K remains at zero, and it then knows that all the numbers are in the right order.

Operation 6

This operation just prints out the contents of the array called X in order i.e. X(1), X(2) . . . X(N). Since the numbers in those boxes are now in ascending order the printout will print the numbers in the right order.

The reader will no doubt consider this to be a rather cumbersome procedure but the speed at which the machine undertakes the task is exceptionally fast and with long lists it can easily out-perform a human being.

EXAMPLE 3

This example (see fig. 5.4) relates to program 2, written and described in chapter 6, for calculating whether it is worth installing double glazing from an economic point of view. It contains within it a routine for sensitivity analysis to discover how sensitive are the results that you obtain to a change in the interest rate. The approach to this chart is rather different to example 5.2 as the actual calculations

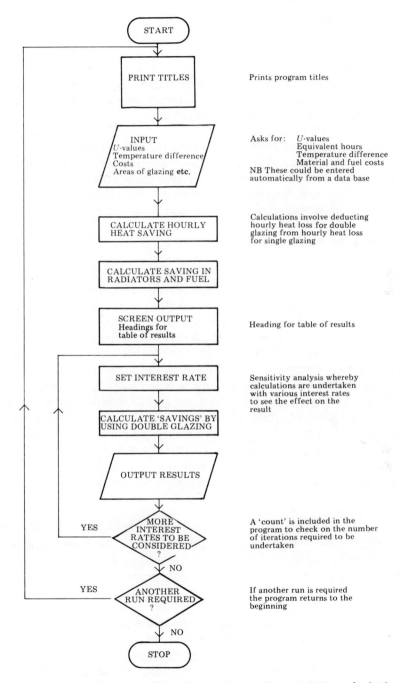

Fig. 5.4 Flow chart for program 2 (see chapter 6): calculation of whether it is economic to install double glazing with sensitivity analysis on interest rate.

are not shown. The diagram merely gives an overview of the procedure without going into the detail of its content. If necessary each operation could be shown in greater detail on a separate sheet but this is not likely to be required providing the programmer knows what computations need to be undertaken. The flow charts for the starter packs at the end of this book are in this form as the programs themselves are available for reference.

As with all planning tools it is wise to remember what they are there for. They are there for the benefit of the planner and should not be a millstone round his neck. Used wisely when and as required they can be helpful; used merely because they exist they can slow down progress and be counter-productive. Experience will dictate to each individual when and how each should be used.

Structuring the program in sections
This has been included as a separate activity merely to emphasise the need to have a reasonably tidy program. Having sorted out the logic, if the various operations can be written in fairly self-contained blocks then it is much easier to insert new routines or provide linkages to other programs or data bases at a later date. It is extremely difficult to develop a program when it has been assembled piecemeal with very little thought for the future. At its simplest the blocking procedure consists of a menu which identifies the options in the program and which the operator chooses to access. Each option will be independent of the other and therefore self-contained in the program listing, making it much easier to read. However most programs consist of a series of interrelated parts and it is here that the skill of the programmer is used to identify self-contained routines which are then linked using other instructions. Again experience will determine the size, quantity and arrangement of program sections.

Write the program
Since the whole of section II is concerned with this aspect we shall not cover it here. Suffice it to say that providing the planning of the program, including the logic and routines, has been well thought out there is no reason why this should not be a straightforward exercise.

Debug the program
Getting rid of errors in programs can be the most frustrating task in the whole of computing. Some slight error in keying in a program or an incorrect line number in an instruction can cause enormous difficulties. There have been very few large programs written in which a 'bug' has not been found. It is in the nature of human activity that mistakes are made and in computing that can be dangerous.

Computer errors in the United States Defense Department have already produced unnecessary 'red alerts' and yet the security and testing of their models must be very high. It is therefore extremely likely that there are many errors even in programs which are in constant commercial use today.

By comparison, at the rather mundane level of building appraisal it is almost certain that we shall find errors — consequently we need to adopt procedures for coping with them. In most of our programs it will be just a case of comparing the machine's results with a manual calculation on the same data. Logic will also need to be tested to ensure that the right routines are being picked up when choices are being made. When errors do occur it is necessary to systematically discover the part of the program where the fault is occurring. In difficult cases this may involve putting extra temporary lines into the program to output interim results to find the point and cause of the error.

An increasing difficulty, owing to the growth of simulation programs involving random numbers, is how to check a program when the data it is using has been generated by chance. The usual method is to artificially fix the numbers used to check that the calculations are correct. After a manual check it is then possible to revert to random generation (ensuring that this generation is actually happening) and allow the machine to produce results.

The checking process in computer programming is rather like a detective solving a crime. The programmer looks for clues which suggest to him that something is going wrong at a particular point. He then narrows down his field of search until the guilty culprit, possibly a line or lines of program, is exposed. The only difference is that in programming it is usually the detective who is responsible for the crime in the first place!

Program implementation

Once the program has been written and debugged it is not necessarily the end of the story. We still need to persuade the office staff to use it, and, through the process of presenting the finished work to someone with no previous knowledge of its structure, we find out where the operator's difficulties lie. In most programs this will inevitably mean a period of refinement, where some of the following may take place:

1. Improvement of screen/printer layout.
2. Additional checks on input of information.
3. Additional explanation of how to input information.

4. More 'help' routines to which the operator can refer should he find that he can't understand his next move.
5. Improved titling of programs. This is usually left until the end to see how much memory store is left. It is surprising how a good title routine (perhaps involving animation) can gain the confidence of the user and improve the public relations image of the firm.
6. Additional options not previously thought of or considered worthwhile.
7. Improved security of information.

It has previously been mentioned in the book that staff training is essential. Nothing is worse for the credibility of appraisal programs if the staff do not fully understand how to use them. Careful instruction overcomes this problem and is well worth the time involved.

Structuring files and data

So far we have looked at the general problem structure and the processes that need to be gone through in order to obtain a satisfactory program. In this general discussion we have played down the problem of 'the system'. We have not discussed how one program may interact with another — which usually means how it provides and receives data from another program. Almost inevitably this will be through the medium of the data base. Calculations from one series of operations are output to a file which is then accessed by another program which uses that information for its own requirements. In concept this is a fairly straightforward procedure as the data is pushed out to some form of backing store, e.g. disk or tape, and then brought back in again by a search of that store. The mechanism used for this transaction is *the file* and this forms the basis of the data base.

Computer files

Files can be considered to be the framework around which data handling or processing revolves. The term 'file' is used to describe a collection of related records rather like the files in the office, such as job files, technical index etc., but this time held in electronic form in the computer. It holds data which is required for providing information, some of which will be processed at regular intervals (e.g. names of suppliers) and some of which will be processed irregularly (e.g. a file containing the price of items).

A file consists of a number of *records* and each record is made up of a number of *fields*. Each *field* then consists of a number of characters.

An example of a record was included in chapter 4, on page 90. Since it is the speed at which each record can be accessed which gives credibility to the program and makes it more efficient, this is obviously an important factor in structuring the problem. It is a difficult area as it nearly always involves compromise. In many respects it may be sensible to place into one record all the attributes relating to a building or component. Apart from the fact that there are limits to how much information can be held in one record there is also the problem that because only a small number of those attributes may be regularly accessed you may be wasting valuable program time — in order to access a relevant piece of information. It may therefore be much more sensible to include the attributes most commonly used in a separate file rather than get the machine to obtain and break down a comprehensive record each time it wants that piece of information. Another alternative is to have a program which will take the comprehensive record and produce a file containing the required information at the request of the operator or the program to avoid permanently holding a number of small files. It need only do this when specifically requested to do so and therefore save on memory store.

Data storage and retrieval
In addition to the length of records and the amount of information held in files a major factor in the speed of operation and compactness of the data base is the manner in which the information is held and retrieved. Some of the questions affecting the choice of file storage are listed below:

1. Is speed of input or retrieval from store of prime concern to the user? If it isn't then simple approaches are probably the best.
2. Do a large number of relationships exist between the components of the system which need to be identified, e.g. the thermal properties of cladding will influence heat loss and therefore installation and running costs of heating systems and even the structure? If these relationships do exist on a large scale then the data base must include for cross-referencing between items of stored data.
3. Is the computer system large enough (in terms of memory capacity) to handle and store a series of complex relationships?
4. Will the system be operated by a skilled or an unskilled casual operator? In the latter case simple routines are preferable or advanced user programs need to be written to lead the operator gently through the system.
5. Will the system be extended in the future? If it will be, then a

system which allows further information, programs etc. to be fed in without wastage of memory store or rewriting of the operating programs is required.

6. How often will the system information be updated? If regularly, then rapid storage and retrieval techniques are required.

These are just a few of the considerations that need to be borne in mind when setting up a data system and the results of deliberation on these points could result in a very complex arrangement of data. It would be inappropriate to deal with such complexity in a book of this nature and therefore only simple arrangements have been included.

File arrangements

SERIAL FILES

This is when records are written on to tape or disk one after the other without there being any relationship between the records, i.e. no regard is paid to the *sequence* of the records. It would be rather like filing all your technical data into a loose leaf folder with no regard to order, classification, date or any other factor.

It has of course the advantage of being very simple but it does mean that the only way to find a piece of information is to go methodically through every item until you find it. This is very time consuming but it does not waste memory space in storage because any additional record is merely added onto the end of what was previously input. In diagrammatic form a serial file is as follows:

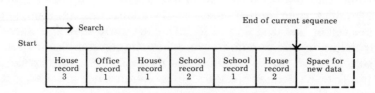

No strategy can be employed to find a piece of information and therefore the computer has to laboriously go through each item until it finds it. However the arrangement is quite often used for preliminary storing of information. Sort programs are then used to access this shuffled pack of data to get it into some useful order. For example:

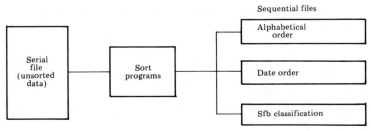

This has the advantage that new information can constantly be fed into the serial file and a full sorted list provided on the total information available without wastage of memory. The ordered list can now be used by other programs. Although memory is not wasted in having to leave gaps for future additions, it does mean that more space (or more floppy disks) are required to hold the sort program and the new structured lists.

SEQUENTIAL FILES

In these files the items included in the system are ordered in some way (e.g. ascending or descending numerical sequence or cost, alphabetical sequence etc.). Data is placed 'end to end' and therefore to insert data or interrogate the data base the beginning and end of the current sequence must be known. In diagrammatic form a sequential file is as follows:

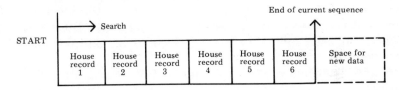

It would be very difficult in this system to add additional records which might need to go between previous records. The order must be known in advance and therefore it is suitable for data which is updated infrequently, e.g. a telephone directory or dictionary.

It is possible to sectionalise the data and leave spaces for new additions:

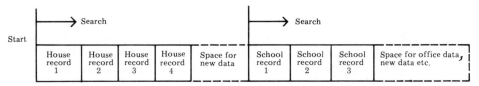

It does mean that each section must be of a stated finite length, and to avoid either insufficient space or wasted storage space, a good idea of the quantity of data in each section must be known.

Once again the system is simple but is only really suitable where the input of data is relatively static.

The fact that the data has already been sorted does mean that some search strategies can be employed which will allow the average access time to be reduced.

Such a strategy is the binary split routine. This method finds the centre item of the ordered data, discovers in which of the two halves the sought item is to be found and then divides this half into two in the same way. The process continues until the required item is isolated. For example if we were looking for the item B in an ordered list of eight items to find out what the coded item B contained, the process would look like this:

Split 1	A	B	C	D	E	F	G	H
Split 2	A	B	C	D				
Split 3	A	B						
Split 4		B						

This technique reduces the average search time (t) from being proportional to the length of the list ($t \propto n$) to being proportional to the log to the base of the number of sections in each split multiplied by the number of items in the list ($t \propto \log_2 n$). t is the average search time and n the number of items. In a binary split the number of sections in each split is two but this could be expanded in larger data bases.

Obviously searches of this nature are not really suitable for tape storage because most tapes will only search in one direction.

Sequential files involving search routines require each record and the file itself to be 'indexed'. This index allows a particular record to be identified quickly for access purposes.

Another variation is the 'selective sequential' file which as the name suggests selects only those items from a serial file which are appropriate to the transactions which are to be undertaken. Sort routines can then be applied to this selected group of records.

RANDOM FILES

In this type of file the records are placed onto the disk 'at random'. At first glance there would appear to be no obvious relationship between the records as there is with the sequential files. What happens is that as each record is keyed in the machine applies a mathematical formula which generates as an answer a 'bucket' address and the record is then placed on to the disk at that address. The 'bucket' refers to a group of 'blocks' (these are the smallest addressable part of a disk) to form a larger unit of transfer. The size of the block or bucket will be optimised by the manufacturer bearing in mind such factors as, 'How do we obtain the minimum number of transfers between main storage and disk?', and, 'How do we economise in the use of main storage?'

In fact there can be a relationship between the records in the file and this can be achieved by pointers which direct the computer from one record address to another. This is known as 'linked storage' and is really beyond the scope of this book. However it is important as it leads to hierarchical data structures which can be of enormous benefit in improving the speed and efficiency of access to data. The concept of simple hierarchies has already been introduced in chapter 4 to show how they are used to get into a system to undertake a particular appraisal. The same principle applies in data structures, only the objective is now to obtain a particular piece of data rather than a program.

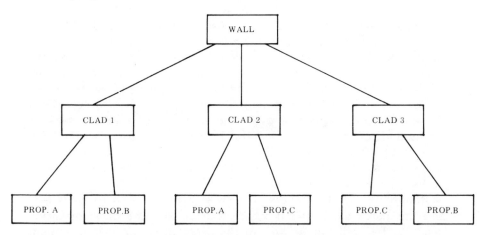

Fig. 5.5 Hierarchical structure for wall claddings.

Fig. 5.5 shows a simple hierarchical structure (which would be set up by the index in the addressing of the records) for wall claddings. With this structure it would be very easy to discover what were the properties of cladding 1 (e.g. thermal, acoustic, cost characteristics) but more difficult to discover which claddings contained the property B. To find the latter we would have to search every property record. If this was to be the main method of interrogating the data base then it may be more appropriate to reverse the hierarchy, as in fig. 5.6. This very simple example illustrates just one of the many

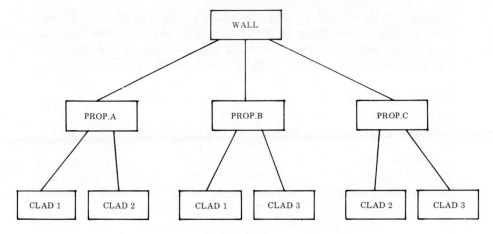

Fig. 5.6 Reversal of structure in fig. 5.5

sometimes difficult decisions that have to be made in order to achieve a fast and efficient information structure.

Of course in practice the hierarchy for a complete library of building components would be extensive and the speed at which the computer can search up and down the system would be critical. The more sophisticated systems do allow a variety of ways to enter the search routine, i.e. at various levels, and to move from the bottom of the tree up the system as well as the more conventional approach.

The more flexible the system the faster the speed of access but the greater the number of overheads in terms of memory requirements and programming. A careful balance needs to be worked out in each case depending on the size of the data store, the number of times items are input and retrieved, the memory store and the importance of increased speed.

Random-access files allow these search routines to be developed and are particularly useful where a large amount of information needs to be available for use in the program. They have the major advantage that unlike the sequential files there is no need to waste memory space in order to accommodate further data and above all each item or record can be accessed direct (usually through an index to the record) and thus avoid a prolonged search process.

Appropriateness of file structures to building appraisal
There can be no hard and fast rules as to which file structure to adopt in appraisal techniques.

Obviously if there is very little data to be accessed or if it is updated infrequently (and this might apply to standard component attributes such as thermal values) then a sequential file is probably satisfactory. If there is a considerable amount of information being handled and it is regularly being added to, then a serial file with program sort routines may be most appropriate. By keeping a master file of current unstructured data, the basic information is kept up to date; it is then possible for any ordering to be separately undertaken knowing that the sequential file produced will also be up to date. No wastage of storage will have occurred. This technique may be suitable for holding the names of suppliers or contractors and possibly elemental analyses.

Random-access files will be useful when it is necessary to obtain a single record quickly and either use it in a program or update it. By having the facility to access an item direct, the search process is reduced considerably. Therefore if the appraisal program picks out items of data in an unstructured way or requires updating in an *ad hoc* manner then random access is extremely useful. It also tends to save on main memory store.

The whole question of file handling and data structure is one that is interesting but complex. Chapter 9 gives an introduction to the mechanics of file handling as far as programming is concerned. It also introduces the concept of 'relative data files' where all data records are of the same length. This is merely a form of standardisation which allows a manufacturer to simplify his file handling system. Unfortunately it is not possible to dictate a standard approach for all machines. Each manufacturer will have developed his own system to suit his own requirements. This lack of common methodology will mean that the operation of a particular machine will have to be learnt from the manual although the principles outlined in the book will apply.

Conclusion

The purpose of section I has been to introduce the reader to some basic concepts in computing. Each aspect is worthy of further study and a list of further reading is included in Appendix I.

Section II consists of an introduction to computer programming through the medium of example programs in building appraisal.

SECTION II
BASIC Programming Related to Building Problems

CHAPTER 6

Introducing BASIC — solving simple formulae

In the first section of the book we looked at the way in which building appraisal problems can be structured and solved by the microcomputer. We also introduced the idea of a computer language which would enable us to overcome the difficulties of communicating with the machine without the need to write our programs in binary form. Over the years many languages have been developed for this purpose such as FORTRAN (FORmula TRANslator), ALGOL (ALGOrithmic Language) and COBOL (COmmon Business Oriented Language) each of which was geared towards a specific application. However the majority of microprocessors use a language called BASIC developed in 1964 by Kemeny and Kurtz at Dartmouth College, New Hampshire, USA. BASIC stands for Beginners All-purpose Symbolic Instruction Code and as the title suggests it is one of the simplest of the high level interactive computer languages. The language has proved so popular that many microcomputer manufacturers have designed their machines with BASIC hardwired into the computer. Since 1964 it has undergone a number of extensions by manufacturers, colleges and universities so that in the majority of systems it is a very powerful tool indeed.

The popularity of BASIC is undoubtedly due to its close resemblance to commands found in normal speech and the ease with which it can be learnt and remembered. In fact it is possible to grasp the essentials in a few hours of practice on a machine.

The purpose of this section of the book is to introduce the reader to various BASIC statements and the way they can be used to help solve building problems. However it is hoped that by the end of section II the reader will be able to take the skeleton programs introduced in the text and those in the 'starter pack' programs and develop them for his or her own particular needs.

This brings us to the method of learning. There is no doubt that the best method of understanding the language is to practice with the machine in front of you. By typing in the programs as they are introduced in the text you will find that you will soon become familiar with the commands, syntax, layout and format of the language. As with spoken languages it is practice that produces fluency and improves vocabulary. Very few people would consider reading a French/English dictionary through from cover to cover and then attempt to speak French to a Frenchman. In the same way it is not advisable to try to learn all the different expressions in a computer language before trying them out on a machine. The knowledge should be built up little by little until confidence is gained in the structure and use of the terms and vocabulary.

Although most of the common BASIC statements are explained, it is not intended that this section of the book should be a comprehensive introduction to the language. There are many such volumes on the market, often geared to a particular make of computer. The chapters in this book should place the learning of BASIC into a building context through the use and understanding of the programs provided. Do not worry if you do not fully understand all the implications of each instruction. Try the programs out by keying them in and typing RUN followed by touching the RETURN or EXECUTE key. Gradually, as you see what the program does and by referring to the text for clarification, you will build up your knowledge of your machine and of the BASIC language. Take time over your progress and thoroughly learn one section before moving on to the next.

There is just one word of warning. BASIC is continually developing and the instructions found suitable for one manufacturer's machine may well be different from another. However this usually applies only to the more advanced instructions which the makers have introduced to improve the power of their particular product. The original statements found in early BASIC tend to remain the same and so does the layout of the program. To overcome this particular problem an attempt has been made in this book to use only those instructions in common use and where differences are likely to occur these will be identified in the text. In addition, it should be mentioned that there are good and bad ways of laying out the program in order to obtain the most efficient use of the machine's memory and also to achieve the maximum speed of operation. While this has been borne in mind in the programs which follow, the prime consideration has been the ease of learning. For this reason clarity has been placed before efficiency of operation. The final chapter of this section does, however, give some useful tips to assist in developing your programming skill.

A simple program

We have already seen in chapter 5 that a program may be considered to have three parts:

(i) Input data: i.e. the information necessary for running the
     program.

(ii) Processing: i.e. the manner in which the input data is to be manipulated.

(iii) Output information: i.e. the information required to be presented from the results of the processing.

To demonstrate these operations let us take the example of solving a simple formula to find the optimum number of storeys of a building when considering only the envelope costs. The formula given by Ferry and Brandon in *Cost Planning of Buildings* (Granada, 1980) is:

$$N^2 = \frac{xf}{2ws}$$

where N is the optimum number of storeys, x is the roof/wall cost ratio, w is the width of building, s is the storey height of building and f is the gross floor area.

The solving of a formula of this nature is perhaps the simplest type of building problem. However we can still apply the general approach that was also outlined in chapter 5 to its solution. Let us summarise the various stages again in relation to this particular case.

1. *Define the problem.* To solve the equation for optimum number of storeys, given the value of the other factors in the equation, and to test the sensitivity of results to changes in the roof/wall cost ratio.

2. *Where necessary provide a mathematical expression for the problem.* In this case the formula has already been given.

3. *Determine the kind of output required.* For this program a table which gives a list of the results of the optimum number of storeys for changing values of the roof to wall cost ratio will be suitable. (In larger programs draft layouts should be considered before writing the program.)

4. *Determine the inputs required to achieve the above output.* In this case we shall need:

    (i)   width of building
    (ii) storey height of building
    (iii) floor area of building
    (iv) various values of roof : wall cost ratio.

5. *Design the program logic to produce the desired results and output.* See flow chart, fig. 6.1.

6. *Structure the program in sections.* In this simple case we can divide the program into three parts as suggested earlier:

    (i)   Request for input information.
    (ii) Calculation of results.
    (iii) Printing of results.

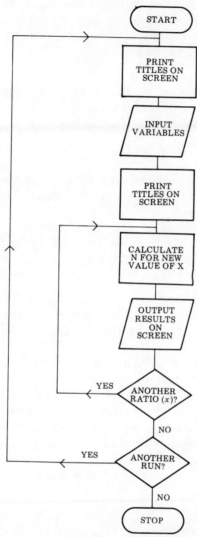

Fig. 6.1 Flow chart for program 1: finding the optimum number N of storeys of a building for changing roof/wall cost ratio x.

7. *Write the program.* Although the BASIC instructions have not yet been introduced you may find it helpful to see how much of program 1 (on page 135) you can already understand. Many of the statements correspond to instructions that would be given in English, e.g. PRINT and INPUT. Even the numerical expressions are recognisable; line 130 contains the formula we are trying to solve.

Program 1 Optimum number of storeys (Dr D. J. Ferry).

Program

```
10 REM OPTIMUM NUMBER OF STORIES

15 PRINT CHR$(147)
20 PRINT "  OPTIMUM NUMBER OF STORIES : DR FERRY"
30 PRINT

40 PRINT "ENTER THE FOLLOWING:-"
50 PRINT
60 INPUT "FLOOR AREA       ";F
65 PRINT
70 INPUT "BUILDING WIDTH   ";W
75 PRINT
80 INPUT "STOREY HEIGHT    ";S
90 K=1
95 PRINT CHR$(147)
100 PRINT "OPTIMUM NO OF STORIES"
110 PRINT
115 PRINT
116 PRINT "    NO","RATIO","STORIES"
120 FOR X=0.4 TO 2.0 STEP 0.2

130 N=SQR((X*F)/(2*W*S))

134 N=N+0.5
135 N=INT(N)
140 PRINT K,X,N
145 PRINT
150 K=K+1
160 NEXT X

170 PRINT "DO YOU WANT ANOTHER RUN";

180 INPUT A$
190 IF A$="YES" THEN 10

195 PRINT "GOODBYE-PROGRAM FINISHED"
200 END
```

Operation

10 REM statement is just a remark and has no effect in program.
15 Clears screen.
20 Prints title.
30 Leaves a blank line under title. (Lines 50, 65, 75, 110, 115, 145 similar.)
40–80 Asks the operator to input the values of the variables in the equation (except for roof/wall cost ratio) and assigns the input values to F, W and S respectively.

90 Sets counter K to 1.
95 Clears screen.
100 Prints title for results.

116 Prints headings for results columns.
120 Changes value of roof/wall cost ratio for each iteration (loop).
130 Calculates N for current value of wall/floor ratio and input value.
134 & 135 Calculates integer value for N.

140 Prints numerical values of these variables.
150 Increases value of counter K by one.
160 Signals end of iteration commenced in line 120.
170 Asks whether a complete rerun is required.
180 Assigns answer of YES or NO to A$.
190 Checks A$ to see if it is YES. If affirmed the run is commenced again from line 10.
195 If A$ is other than YES the program is completed by this PRINT statement.
200 Stops the program.

The annotation on the right hand side of the page gives a brief explanation of what is actually happening as the machine acts upon the instructions which have been given.

8. *Debug the program.* Getting rid of errors in the program is a time consuming task which everyone faces once they have written and typed a program into the machine.

Program layout

We have now outlined our approach to the problem and need to explain in more detail the BASIC instructions used in the simple program 1. For ease of reference we shall introduce the instructions in the order that they appear in the listing given.

One important point to notice in BASIC programs is that every line is numbered. This arranges the program in order and identifies where a routine is to be found so that the machine can operate efficiently and process the lines in the correct sequence. It is common practice to leave gaps of five or ten between each line, in order that further statements may be added at a later date. However, where lines must follow each other, e.g. a continuing PRINT statement, this practice is unnecessary.

When developing complex programs it is as well to think ahead and build the program up in 'blocks' leaving many more lines between the blocks for further development or for any linking titles, instructions etc. The flow chart should identify the structure of the blocks and the overall program organisation.

To enter a program into the machine it is usual to type the number of the line followed by the program statements and then to press the RETURN or similar key to instruct the machine to place the line in its memory. The finished program can be saved for future use, in most machines, by recording on a cassette tape or disk. The precise instructions for your particular microcomputer will be found in the manufacturer's handbook and must be followed with complete accuracy. The same will apply to listing and editing the program. These instructions vary considerably from one machine to another and an evening learning the particular requirements of your computer will pay dividends at a later date.

REM statements

In any fairly complex piece of writing or calculation it is advisable to use headings and side notes to clarify what is actually taking place. This is the sort of discipline imposed by school teachers when teaching mathematics and by cost advisers when preparing their measurements. In programming it is even more essential that the structure of the program is clearly seen so that the writer knows exactly what each part of the program is doing should he require to revise, add to or correct it. Therefore whenever the end of one operational sequence is reached it is advisable to start the new sequence with a 'signpost' to signify and identify the start of something new.

REM is the statement that allows you to provide these 'signposts'. It is short for the word REMark and this is what it allows you to do. Any phrase, code, sentence or title may be placed after this statement. When the computer comes across a line starting with REM is will either ignore the entire line, or ignore the statement beginning with REM and proceed to the next statement or line. Most of the microcomputers ignore the whole line, however many statements it contains. However, a REM statement does take up valuable memory space so it is advisable to keep the remark as concise as possible. Sometimes programmers use asterisks at the end or within a line to enable the program sections to be easily picked out when reading through a listing.

PRINT statements

This is one of the most useful of all BASIC statements because it allows any text such as calculated values, questions, output of results etc. to be displayed on the screen or output to a printer.

In lines 20, 100 and 195 the PRINT instruction is used to provide a title on the screen, e.g.

20 PRINT "   OPTIMUM NO OF STOREYS : DR FERRY"

The inverted commas after the instruction signify to the machine that it is required to print everything found after the inverted commas exactly as it is shown (including any spaces), until another set of inverted commas is found. So if we require to place the title in the centre of the screen, the space bar on the keyboard can be used to push the title along from the left hand edge of the screen until it is central. The number of spaces needed will depend on the screen size and if a 40-character screen is being used, the above title which is 32 characters long will require four spaces before the beginning of the text. A notation which is sometimes used to indicate a space is an inverted V or U though of course these are not used in the actual program. The term 'inverted commas' has been used above, but the BASIC language is American and the term 'double quotes' is more usual and will be used from now on.

This approach has been taken a stage further in line 116 where the headings to the columns in the final results table need to be printed.

116 PRINT "   NO" , "RATIO" , "STOREYS"

Here commas have been used to space the words out across the screen. The machine assumes a 'field' of a certain number of characters between each comma (usually a field width of 10 is used on a 40-character wide screen). It then prints the word or number in that field, starting from the left hand side. Once again spaces are allowed between the double quotes to centralise a heading within the 'field' depending on the length of the word.

To print the results under each heading the same technique is used but when printing out the values of the variables the double quotes are not required, as in line 140.

140 PRINT K,X,N

Because no double quotes have been included the machine understands that the values held in stores K, X and N (see later) are to be printed. These will be numerical values and these will change each time the calculation is repeated and different values entered into the equation. Once again the comma between each letter tells the machine to set out the numbers in the columns within the field.

Care must be taken when using commas to make sure that either the word or the number does not exceed the width of the field. If it does, then the next word or number will commence an extra field width away from that originally intended. This can completely upset the appearance of the table. Another point to remember, is that if numbers are being printed in these fields, the decimal points will only line up if the digits contained in each number are the same.

Lines 30, 50, 65, 110, 115 and 145 contain a PRINT statement with nothing following it. This is the method which is used to obtain a blank line on the screen and so make the text and numbers easier to read. In some machines a double set of double quotes is required after the word PRINT to gain the same effect.

The final use of PRINT in this program is in lines 40 and 170 where it is used to give instructions to the operator or to ask the operator a question. In both cases it is used with an INPUT instruction and this is covered under the next heading.

INPUT statements and variables
To be able to solve an algebraic expression it is of course necessary to give numerical values to the letters contained in the expression. There are several ways of doing this but in this chapter we will consider only two methods. The simplest way to give a numerical value to a variable is as follows:

90 LET K=1

or the shortened version of this which is available in most modern versions of BASIC:

90 K=1

It is important to understand what this really means. The K in the expression simply means that it is the address or location of a store which is to be called K. Rather like a bank of pigeon-holes for filing, the machine has to pick a store and it will be referred to as K. The expression K=1 means that the figure 1 is to be placed in this store. If we then add, subtract, multiply or divide using K in an expression, the machine will always take the *contents* of the store K and use this in the arithmetic operation. It has *assigned* a value to store K which it will continue to use until it is told that this value has to be changed.

If we look at line 150 we can see an instruction which changes the value of K:

150 K=K+1

This expression would be nonsense in algebra but is allowable in BASIC. What it means is that the current value of the store K is to have one added to it and the result is to replace the old value of store K.

This principle applies to all cases of stores which contain numerical values. It therefore means that the left hand side of any statement (i.e. to the left of the equals sign) can only contain the address of one store. Thus an expression such as:

150 K+1=K

is not possible; it must be:

150 K=K+1

In line 150 we are keeping an account of the number of times a routine within the program is processed. Each time the calculation is repeated, one is added to K.

When values are included in a program in this manner, the operator has no choice, without altering the program statements, as to what the value of K should be. Its value changes as specified.

We must also have a means of entering the variable value using the keyboard while the program is being run and assigning that value to a store. The contents of the store can then be manipulated in an algebraic expression. The way in which this can be done is through the INPUT instruction. Let us take line 60:

60 INPUT "FLOOR AREA   ";F

This instruction has the following effect:

(i) It stops the program at line 60.

(ii) It prints FLOOR AREA and six spaces, i.e. the literal characters it finds between the double quotes, followed by a question mark.

(iii) It expects to receive from the operator a numerical value which it will assign to a store called F. The semicolon (some versions of BASIC require a comma here) means that the prompt for F (often the flashing cursor) will occur just after the sixth space.

Until the operator has entered a value and touched the RETURN key the program will not continue. The number which has been entered is then placed in store F and held until it is used in the calculation of the optimum number of storeys in line 130.

NB It is, however, important to realise that this method of using INPUT is not universal to all machines. It may be necessary to split the instruction into two lines, the first to print the words and the second to input the value of F in the following manner:

```
60 PRINT "FLOOR AREA        ";
61 INPUT F
```

In this case the semicolon at the end of line 60 tells the machine that the next output to the screen (in this case the prompt to enter the value of F) is to be printed on the same line as the text contained in the PRINT instruction. If your microcomputer is found to require this approach then lines 60, 70 and 80 should be altered accordingly.

There is one final use of the INPUT instruction in the program and this is where the response expected by the machine is a word instead of a number:

```
170 PRINT "DO YOU WANT ANOTHER RUN";
180 INPUT A$
```

INPUT in this context performs in the same way as it did when a numerical value was required. The difference is that the store for the variable contains a dollar sign as the last character. This dollar sign signifies to the machine that the input will be of an alphanumeric character, i.e. it may contain all letters, all numbers or a mixture of the two. Therefore all the stores in the computer which have been labelled with the suffix $ will contain information which is to be treated as non-numeric. They cannot be manipulated (e.g. multiplied together) in the same way as numeric stores even if they should contain all numbers. The variables of this type are known as 'string' or 'alphanumeric' variables and are covered in more detail in the next chapter. In this program the word expected as an answer is either YES or NO in response to the question "DO YOU WANT

ANOTHER RUN". Since we only want to take any action if the answer to the question is YES, we only need to check that the word input into the store A$ is in fact YES. This is achieved by comparing the store A$ with the word we are looking for.

190 IF A$="YES" THEN 10

What this line does in the program is to:

(i) Call up the contents of the store A$.
(ii) Compare the contents with the word in double quotes, in this case YES.
(iii) If the comparison proves that the contents of A$ and the word YES are identical it tells the machine to start the program again at line 10. If the input word is proved not to be YES then the program continues and prints the GOODBYE slogan.

In some cases it is desirable to check both possible answers and make the machine respond only to either YES or NO. This has introduced another BASIC statement, IF. These IF statements are dealt with in more detail in chapter 8 but it can be seen from this example that simple decisions, where the answer is YES or NO, can be taken according to the operator's wishes.

FOR-TO-NEXT statements
It is often useful in a program to be able to tell the machine to repeat a calculation a given number of times and on each repeat to change the value of one of the variables by a specified sum. It is also helpful if this can be done automatically without interference by the operator.

Lines 120 and 160 allow us to do this in the example program.

120 FOR X=0.4 TO 2.0 STEP 0.2

In our notation X is the roof/wall cost ratio. This line instructs the computer to change the value of X from 0.4 to 2.0 inclusive, in steps of 0.2, and to recalculate the value for the optimum number of storeys (N) each time the value of X changes. To tell the machine how long it should retain the new value of X, a NEXT statement is inserted at the point of change.

160 NEXT X

When the machine comes across a NEXT statement it returns to the statement following the FOR-TO statement to which it belongs and changes the value of the variable (in this case X) to the next incremental step. These statements are explained in more detail and developed further in the next chapter.

When using decimal fractions it is common practice to write a zero before the decimal point. In writing a computer program each zero takes up memory space and they are not required by the machine. It is therefore usual to omit these, so that line 120 above would normally be written as:

120 FOR X=.4 TO 2 STEP .2

Arithmetic operations

The statements introduced up to now have largely been involved with the input and output of information. However the really powerful ability of the machine is to perform calculations quickly. The useful part of our example program arithmetically is line 130 which contains the formula. Before explaining the operation of this line, an introduction is required to the symbols used in arithmetic operations and the order in which those operations are performed by the machine.

The symbols are:

+   To add one number to another.
—   To subtract one number from another.
*   To multiply one number by another (avoids confusing a multiplication sign with X).
/   To divide one number by another.
↑   To raise a number to a power.

If we now look at line 130 we see some of these in operation:

130 N=SQR((X*F)/(2*W*S))

You may wonder why there are so many brackets. These are required in certain cases to ensure that the arithmetic operations are undertaken in the correct order. By placing brackets around groups of operations it ensures that the values in the brackets are calculated first and therefore can be treated as single expressions. In this case (X*F) and (2*W*S) will be calculated as single expressions before one is divided by the other. Brackets may of course be nested and are evaluated outward starting from the innermost set. The order of precedence of operators is:

Exponentials	first
*M*ultiplication & *d*ivision	second
*A*ddition & *s*ubtraction	third

You may like to remember this order by the phrase '*M*any *D*rowsy *A*rchitects *S*leep'.

Using the above priorities, all expressions are evaluated from left to right. In our particular expression the first set of inner brackets is not necessary and omitting them would save memory space. Line 130 could therefore be written:

130 N=SQR(X*F/(2*W*F))

If the second set of inner brackets is omitted then the square root is found of X multiplied by F divided by 2 multiplied by W and multiplied by F which would, of course, give the incorrect answer. If the second set of inner brackets is to be omitted the line must be written:

130 N=SQR(X*F/2/W/F)

However, it is always safer to put in too many brackets until you gain sufficient confidence.

If it should happen that the divisor in any expression is equal to zero or if the result of an expression is negative where a square root has to be taken, an error is signalled. Protection can be provided against this by testing the value of W or F to see if either are zero before line 130 is reached. If they are, a message can be printed to say that this has occurred. A test could also be included to test the expression to see if it produces a negative answer before attempting to take the square root. The program can then return to the relevant line to permit re-entry of the data.

Lastly in this particular expression we wish to find the positive square root. In the BASIC language there is a set of standard mathematical functions which are really small program routines to find the value concerned.

Some typical standard functions are listed below:

SQR( ) Calculates the square root of any number greater than zero within the brackets. If the number within the brackets is zero or negative an error will result.

SIN( ) Calculates the sine of the angle within the brackets. The angle is normally assumed to be in radians.

COS( ) Calculates the cosine.

TAN( ) Calculates the tangent.

ABS( ) Returns the value of the expression within the brackets as a positive number.

INT( ) Rounds the value of the expression within the brackets to the next lowest integer (whole number).

In our example program SQR( ) is used in line 130 to find the square root of the expression within the outermost brackets.

The function INT( ) is also used in the program in line 135. Since the optimum number of storeys must be a whole number the result of the expression in line 130 (i.e. N) must be rounded up or down.

Unfortunately the INT( ) function in BASIC will only round down or chop off any figures beyond the decimal point. Therefore an additional line has to be included to make the machine round up or down depending on the figures following the decimal point. In this program line 134 adds 0.5 to the value of N and then the integer value is taken. If for example the value of N was 6.3 and 0.5 added, N would become 6.8. Rounding this to an integer would result in N being set to 6. If the value of N was 6.8 and 0.5 added, the integer figure becomes 7. Hence the difficulty has been overcome.

END statement

This is the final statement in the program and it brings the program to a halt. In many machines this statement is optional and the program halts when the last line of the program is reached.

Some developments

The above statements conclude the explanation of program 1. Even with the few commands given above, it is possible to write a number of simple programs which will be of use in the office of the design team.

Having typed the program into the machine you should of course try running it. Usually this involves typing RUN and touching the RETURN key. Do not be surprised if the program does not run immediately. Common faults that occur are either incorrect typing of the instructions (particularly the number of brackets) or the use of incorrect statements. The latter may be due to the different BASIC language form in your machine. Check your own handbook to ensure that your statements comply with its requirements.

When you have typed in program 1 and succeeded in getting it to run, the results you should obtain for:

a floor area of 5000 m²
a building width of 10 m
a storey height of 4 m

are as follows:

NO.	RATIO	STOREYS
1	0.4	5
2	0.6	6
3	0.8	7
4	1.0	8
5	1.2	9
6	1.4	9

7	1.6	10
8	1.8	11
9	2.0	11

Note: in some machines the program may only run up to loop number 8 (X=1.8) due to an error in the decimal point addition. In these cases merely change line 120 to:

120 FOR X=0.4 TO 2.2 STEP 0.2

When it is running successfully you can then try some editing of the program. Here are two suggestions:

1. Change the values in line 120 to try a different range of options for the roof to wall cost ratio.
2. Enter new lines into the program which will ask the operator for the job name and date and include these new values in the printout of the final results table.

The editing facilities (usually involving the cursor) will vary from machine to machine. Your handbook will give details which you should learn.

To review your program on the screen or printer it is usually sufficient to type LIST and then touch the RETURN key.

Glazing evaluation program

Once you have become familiar with the above program and your machine controls you may like to type in program 2 on page 147. This program discovers whether it is economic or not to install double glazing in a new building. It is based on the method outlined by Bathurst and Butler in *Building Cost Control Techniques and Economics* (page 74; Heinemann). All the BASIC statements have been introduced in program 1 and the two different methods of using INPUT are used to show how each can be adopted (provided that your machine allows you to do so). The flow chart for this program is shown in fig. 5.4.

The method adopted for evaluating double glazing uses 'equivalent hours' and operates as follows:

1. Calculate rate of heat loss through both single and double glazing and subtract one from the other to obtain net heat saving in watts (lines 250—270).
2. Calculate radiator saving by dividing radiator output per square metre of heating surface into the heat saving (line 320) and multiplying by the unit cost of the radiator per square metre of heating surface.

3. Calculate annual fuel saving by multiplying rate of heat saving by the 'equivalent hours' (obtained from a table) which is converted to kilowatts by dividing by 1000 and multiplied by the cost per useful kilowatt hour of the fuel (line 340).

4. The present value of the annual fuel cost is found (lines 430–450) by multiplying by the present value of one pound per annum using the year's purchase formula. This saving together with the radiator saving is deducted from the extra cost of double glazing (line 460) to give the net saving or otherwise of the installation over a 60-year building life.

A sensitivity analysis is included in the program so that the effect of changing interest rate on the final result can be seen.

Taking the following values:

Q1	Area to be double glazed	750	m^2
Q2	U-value for single glazing	5.35	W/m^2 °C
Q3	U-value for double glazing	3.15	W/m^2 °C
Q4	Equivalent hours	2147	h
Q5	Maximum temperature difference	16	°C
Q6	Radiator cost per square metre	25.00	£
Q7	Extra over cost of double glazing	30.00	£
Q8	Cost per useful kWh of fuel	0.0072	£
Q9	Radiator output per square metre	500	W

The results you should obtain are:

INT. RATE	FUEL SAVING	COST SAVING
1	18346	−2834
4	9233	−11947
7	5729	−15451
10	4068	−17112
13	3137	−18043
16	2550	−18630
19	2148	−19032

The only other operation in these programs which has not already been covered is the 'clear screen' instruction found in lines 40, 190 and 360 of program 2 and lines 15 and 95 of program 1. This again will vary from one machine to another and the handbook will give the method that applies to your machine. In the Commodore PET, for example, the clear screen instruction is press SHIFT and the CLR/HOME key at the same time and a reversed heart is printed on the program line. Alternatively PRINT CHR$(147) produces the same result and has been used in all the sample programs. The effect is to wipe the screen clear and move the cursor to the top

Program 2 Comparison of single and double glazing.

Program

```
40 PRINT CHR$(147)
50 PRINT "EVALUATION OF SINGLE V DOUBLE GLAZING"
60 PRINT

70 PRINT "ANSWER THE FOLLOWING QUESTIONS:-"
80 PRINT
90 PRINT "AREA TO BE DOUBLE GLAZED(SQ. M)"
100 INPUT A

110 PRINT "AVERAGE U VALUE FOR SINGLE GLAZING"
120 INPUT "(W/SQ.M DEG C)";U1
130 PRINT "AVERAGE U VALUE FOR DOUBLE GLAZING"
140 INPUT "(W/SQ.M DEG C)";U2
150 PRINT "EQUIVALENT HOURS FOR YOUR BUILDING"
160 INPUT E
170 PRINT "MAXIMUM TEMPERATURE DIFFERENCE"
180 INPUT "(DEG C)";T
190 PRINT CHR$(147)
200 PRINT "ENTER COSTS OF THE FOLLOWING (POUNDS):-"
205 PRINT
210 INPUT "RADIATOR COST PER SQ.M";R1
215 PRINT
220 INPUT "EXTRA OVER COST OF D/GLZG PER SQ.M";G1
225 PRINT
230 INPUT "COST PER USEFUL KW.HOUR OF FUEL";F1

240 REM CALC. RATE OF HEAT LOSS************

250 S=A*U1*T
260 D=A*U2*T
270 H1=S-D

280 G=A*G1

290 REM CALC. RADIATOR SAVING************
300 PRINT
310 INPUT "RADIATOR OUTPUT PER SQ.M.(WATTS)";R2
320 R=(H1/R2)*R1
```

Operation

40 Clears screen.
50 Prints title.
60 Prints blank line. Used throughout program.
70—180 Asks the operator to input the relevant values for the heat loss calculation and assigns the input values to the variables U1, U2, E and T respectively.

190 Clears screen.
200—230 Asks the operator to input the relevant costs for the components and assigns the input values to the variables R1, G1, F1 respectively. Note that this method of inputting information on one line is not available in all forms of BASIC. If not available use the technique shown in lines 70—100.
240 REM statement is merely a signpost when reading the program. Has no effect on operation.
250—260 Calculates heat loss in watts through single and double glazing.
270 Calculates net heat saving in watts with double glazing.
280 Calculates extra cost of double glazing.
300—320 Calculates saving on radiators.

Program 2 Contd.

Program

```
330 REM CALC. ANN. FUEL SAVING**************
340 H=H1*E*F1/1000

350 REM CALC. COST IN USE***************
360 PRINT CHR$(147)
370 PRINT "RESULTS FOR 60 YEAR LIFE OF BUILDING"
380 PRINT
390 PRINT "INT. RATE    FUEL SAVING    COST SAVING"
400 PRINT
410 FOR N=1 TO 19 STEP 3

420 I=N/100
430 Y=((1+I)↑60-1)/(I*(1+I)↑60)
440 F=H*Y
445 F=F+.5
450 F=INT(F)
460 T=(F+R)-G

465 T=T+.5

470 T=INT(T)
480 PRINT N,F,,T

490 PRINT
500 NEXT N

510 PRINT "NB. NO ACCOUNT HAS BEEN TAKEN OF SAVINGS"
520 PRINT "IN BOILER SIZE IN FINAL COLUMN"
530 INPUT "DO YOU WANT ANOTHER RUN";A$
540 IF A$="YES" THEN 10
550 END
```

Operation

340 Calculates annual fuel saving (equivalent hours method)

360 Clears screen
370—400 Prints titles and column headings for results.

410 Changes value of interest rate for each iteration.
420—440 Calculates present value (PV) of fuel saving.

445 & 450 Changes PV to integer value.

460 Calculates total cost saving or loss for double glazing installation.
465 & 470 Changes total to an integer value.

480 Prints numerical value of these variables.

500 Signals end of iteration commenced in line 410.

left hand corner of the screen thus providing a tidy output for any following PRINT statements. It can be omitted without changing any of the program results, merely their presentation. Note that in the PET system the clear instruction where the heart sign is used is given between quotes. This may not be universal. If the second method using CHR$( ) is chosen, no quotes are required. The equivalent command for the Apple (Applesoft) is HOME and that for the Tandy (Level II) is CLS.

CHAPTER 7
A dialogue with data

In all building problems the professional team are involved in handling and manipulating information or data. One of the attributes of the computer is that it can access information very quickly, bring it to the notice of the operator and then manipulate it as required by the program. This chapter sets out some of the BASIC statements that will be used as tools to execute these tasks.

In the previous chapter we made use of the INPUT statement as a method of providing the computer with the information which was needed to run a particular program. The information supplied may well be varied to suit the circumstances each time the program is run.

This chapter will deal in greater detail with the various types of data which may be required by the computer for the successful running of a program.

Information, or data as it is often called, can be divided into two groups:

(a) Data which is constant, in that the same values are used every time the program is run.

10 X=25

sets the variable X to equal 25. Each time this variable is used in a mathematical expression in the program, the machine will automatically use the value 25 in the calculation.

Other examples of this kind of data are the various constants used for the concrete mixes and standard concrete locations and the thicknesses which are set at the beginning of program 3, 'Estimating the price of concrete', included at the end of this chapter (for flow chart see fig. 7.1). The data is allocated to the variables C$( ), M$( ) and P$( ) using the assignment statement. C$( ) is used to contain the mix description and also the quantities of cement and aggregates required for one cubic metre of concrete. The values are set in lines 20—40. M$( ) contains the details of the various concrete mixer types and the values are set in lines 50—80. P$( ) contains the possible locations of the concrete and the labour cost factors associated with these locations. These values are set in lines 90—150.

(b) Data which is fed into the computer by the operator and may be different each time the program is run.

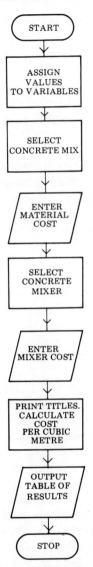

Fig. 7.1 Flow chart for program 3: estimating the price of concrete.

This data is supplied by the operator in response to questions which have been programmed to appear on the screen. If program 3 is being used, the system must be told the current cost of the materials, i.e. cement (line 310), sand (line 320) and aggregate (line 330), in order that it can use the data type (a) described above combined with the new data type (b) and so perform the necessary calculations to obtain the desired estimate.

Any data which is required for the successful running of a program, whether the data is type (a) or (b), must be stored in some location in memory in such a way that it can easily be accessed by the system when it is needed. A simple way of naming these locations is by using the letters of the alphabet. Using single letters will provide us with only 26 names and this small number could well be insufficient for our needs in a program. In order to increase the number of variable names, the letters may be followed by a number from 0 to 9, e.g. A2, Q9 etc. which will provide us with a further 260 names. This gives a total of 286 names we can use for variables which is more than enough, particularly as each name can be used four times in any program, i.e. once for each type of variable (the different types of variable are described later).

In some computers it is also possible to use two or more letters as a variable name. In program 3 the current cost of cement is put into the variable CEM (line 310). It could equally have been put into a variable CEMENT. However, if another variable CENTRE is used in the same program, the system will take this to be the same variable because only the first two letters are taken into account. This facility of using complete words as variables may be useful on occasions, but it must be used with care to avoid confusion.

A further limitation is that the variable name chosen must not be able to be confused with any BASIC word. For example, if a variable is given the name TOTAL, this will be interpreted by the system as the BASIC word TO. A variable named COST will be interpreted as COS etc. In each case this will produce an error.

When choosing a location name it is good practice to select something which suggests the type of data you wish to store — the capability of being able to use two letters can be useful for this purpose. For example, C (CE, or CEM) could be the price of cement; S (SA, SAN or SAND) could be the price of sand etc. Throughout this book upper case letters are used for all variable names and BASIC words. Some machines are now available on which the normal screen display is in lower case, the shift key being required to obtain an upper case letter (as on a typewriter). On such machines all variable names and BASIC words appear in lower case letters.

The values stored in these locations are called variables and in BASIC there are four kinds. These are:

1. Numeric scalar

2. Numeric array

3. Alphanumeric (string) scalar

4. Alphanumeric (string) array

It is necessary to distinguish between numeric and alphanumeric variables as the method of storage in memory is different and their use is governed by different rules.

In the previous chapter, use has been made of numeric and alphanumeric scalar variables using the INPUT statement to assign values. We will now deal with the various types in more detail.

Numeric scalar variables

These are used to store a single numeric value to an accuracy which depends on the amount of memory space allocated by the system to the storage of numbers. This varies with the type of machine being used.

Numeric array variables

These variables consist of a group or array of elements, each of which is in effect a numeric scalar variable. If a group of related variables is required this is a convenient way of storing and accessing them.

For example, the program for the calculation of daylight factors in starter pack 6 requires that the values from a table contained in the UK *Building Research Establishment Digest* 42 be held in memory. There are nine values in this table and the variable used is allocated the name C (which is the same as that used in the *BRE Digest*). In order that it can store all the values it must contain nine elements. The numerical values are therefore stored in the one-dimensional array which is DIMensioned as C(9). This means that the variable C( ) has nine elements C(1), C(2), . . . C(9).

If such variables are to be used, memory space must be reserved for them by the system. As the 'book keeping' involved if each is stored as a separate value may be wasteful, the variable has to have its size specified. The method of doing this is by using a DIMension statement which is in the form DIM C(9) which specifies the size of the array variable C( ). This must appear before the variable is used in the program.

Some microcomputers do not require a small array variable to be specified in this way, the upper limit to size being a one-dimensional array (for the meaning of 'one-dimensional array' see later) with a maximum dimension of ten. There is, however, no objection to dimensioning all arrays and in practice this is preferable.

As we have already seen, in order to execute a program it is necessary to key in RUN and press RETURN. The effect of this is firstly for the program to be 'resolved' and if this procedure is satisfactorily completed, to run the program. Resolving the program means that the system checks through the program to see that it is practicable to run it and also to assign spaces in memory for the variables. Some microcomputers, however, do not carry out this pre-check and errors due to a statement such as GOTO 50 in a program which does not contain a line 50 are not discovered until the program tries to access line 50. All variables in a DIM statement are allotted space at this time, and generally, variables requiring dimensions must be allocated space before variables which do not.

A numeric scalar and a numeric array can have the same name, e.g. in addition to C(9) there may also be a variable C in the same program.

The array C( ) referred to above is called a one-dimensional array, i.e. it consists of a single row of numbers. Two-dimensional arrays are also possible and are designated in the form Q(3,4) which means the array variable has 12 elements arranged in four columns of three rows each as follows:

Q(1,1)   Q(1,2)   Q(1,3)   Q(1,4)
Q(2,1)   Q(2,2)   Q(2,3)   Q(2,4)
Q(3,1)   Q(3,2)   Q(3,3)   Q(3,4)

The limiting size for arrays such as this is normally governed by the memory size of the machine being used.

In the program for starter pack 6, the daylight factors are calculated at each intersection point of a 0.5 m grid in a room and the single variable name R is used for all the results. The array is dimensioned to R(21,21) which limits the size of the room to 10 m square.

In some versions of BASIC, zero is a valid subscript in an array so that the array C(9) would contain ten elements not nine, i.e. C(0), C(1), . . . C(9) and the array R(21,21) would contain 484 elements not 441. As this is not universal, the zero subscript has not been used.

Alphanumeric scalar variables (string variables)
Alphanumeric or string variables may contain either alpha characters or numerals, or a mixture of the two. Mathematical calculations are not permitted on these variables even if they should contain all numbers, although comparisons such as:

=   equals
<   less than
>   greater than
<=   less than or equal
> = greater than or equal
< > not equal

may be used. These signs are known as relational operators. As will be shown later, strings which do contain numbers can be changed into numeric values so that mathematical calculations can be carried out on them. The 'greater than' and 'less than' functions are useful when sorting a list into alphabetical order.

The method of sorting a list in this way is dealt with in the next chapter.

The term 'alpha characters' in this sense includes not only the letters of the alphabet but all the keyboard symbols with the exception of the double quote and the comma (the reason for this is described later).

Any variable which is to contain alpha characters must have the dollar sign following its identification, e.g. the variable A\$ will accept any character. The variable A will accept numbers only.

In the original version of BASIC, the number of characters which could be contained in a string array was normally 16. Modern versions of BASIC usually permit longer strings to be used where necessary and may require the length to be specified in a DIM statement, e.g. DIM A\$25 (the maximum length which can be specified will depend on the machine) allocates space for 25 characters to be held in A\$. It is not possible to specify string lengths on many of the microcomputers and the maximum length of a string is often 255 characters. If an attempt is made to assign more than the specified or permissible number of characters to the variable, the excess will be lost. For example, suppose variable P\$ has not been dimensioned and is only capable of holding 16 characters. The line

30 P\$="FRED SMITH, APPLE STREET"

will result in P\$ being set to "FRED SMITH, APPL", i.e. exactly sixteen characters long.

When values are assigned to string variables, the value must be enclosed in double quotes. It follows therefore, that the double quote cannot be part of a string variable assigned in this way, as it would automatically be interpreted by the system to mean the end of the information to be included in the string.

Alphanumeric or string array variables

In the same way that numeric values can be stored in array form, alphanumeric values can also be stored. The same rule applies — that such arrays must be specified in a DIM statement before they are used.

An alphanumeric scalar and an alphanumeric array may have the same name. In a program which contains the numeric variable C and the numeric array C(9), there can also be a variable C\$ and an array C\$(2,3). Hence each variable name can be used four times in any program.

Methods of supplying the machine with data

The assignment statement
In the previous chapter we used statements such as:

90 LET K=1 (or simply 90 K=1)

to provide the machine with the kind of data which does not have to be changed each time the program is run.

The above example of an assignment statement (LET) sets the variable K to the value of 1. All numeric variables are set to zero when you type in RUN-RETURN and remain so until given some value during the run of the program. In the same way, string variables are set to all spaces or 'nulls' and remain so until other values are assigned to them.

In the case of a numeric variable, the values on the right hand side of the statement need not be a simple number but can be a mathematical expression such as

90 K=P+SIN(P—2.5)

The value of K is then set to the value of the expression P+SIN(P—2.5) and will therefore depend on the value of the variable P at the time this line is processed.

Values may also be given to string variables using the assignment statement. In this case the literal string on the right hand side must be enclosed in double quotes.

100 A$="ABCDE"

gives A$ the value ABCDE or in other words places the letters ABCDE in the store A$.

This method of assigning values to strings is used at the beginning of program 3, 'Estimating the price of concrete', where values are assigned to the elements of the arrays C$( ), M$( ) and P$( ) in lines 20—150.

It is also a legal statement to set one string equal to another.

100 B$=A$

will set B$ to ABCDE (note that double quotes are not used in this case).

Modern BASIC usually permits concatenation of strings, so that:

120 C$=A$+B$

or

120 C$=A$&B$

depending on the BASIC being used will set C$ to ABCDABCD assuming that the previous lines 100 and 110 are part of the same program.

The INPUT statement
This was also referred to in the previous chapter where the statement was used to accept the answers to questions which have been programmed to appear on the screen. Program 3 also requires questions to be answered as the program runs and the statement

240 INPUT"THE NUMBER OF MY SELECTION IS ";Z

is used to assign a value to the variable Z. The value given to Z can vary each time the program is run. Note that some microcomputers require a semicolon between the text and the variable. Other machines require a comma. No question mark need be included within the quotes as the system automatically produces one when it encounters an INPUT statement.

It is also possible to make an INPUT statement ask for more than one piece of information at one time such as:

70 INPUT "BREADTH(M),DEPTH(M)";B,D

Two numeric values are expected as an answer to this question and they are to be separated by commas. The variable B will be set to the first value entered and the variable D set to the second. If only one value is entered before the RETURN key is touched, another question mark will appear on the screen. The program will not continue until both pieces of information have been supplied or the RETURN key has been touched the required number of times. If a system detected error occurs, the values must be re-entered beginning with the erroneous value.

In the above example, if the values entered were Y,10 instead of 6,10 and the RETURN key touched, the system would detect that the wrong type of variable had been entered and an error message would normally be displayed on the screen. It would then be necessary to enter 6,10 correctly and touch RETURN to continue with the program. Touching RETURN twice without entering any values may set B and D to zero or stop program operation, depending on the machine.

An INPUT statement can also be used to accept string values.

140 INPUT "NAME, & ADDRESS";N$,A$

will expect two text strings separated by a comma. It follows therefore that a comma cannot be included in the string.

Answering the question above, J. SMITH, 28 HIGH STREET, PORTSMOUTH will result in N$ being set to J. SMITH and A$ to 28 HIGH STREET. The remainder of the address is ignored as the comma is taken to be the termination of the data. This can quite easily be overcome by answering, J. SMITH, "28 HIGH STREET, PORTSMOUTH". If double quotes are used in this way the whole of the information within the quotes becomes the value of A$. The same method can be used to force leading blanks into a string.

As double quotes are used to define the limits of the string, it follows that these cannot be included in a string in reply to an INPUT statement.

This does not mean that a string can never contain double quotes, simply that they cannot be inserted by the methods so far described.

The READ statement

If a number of variables has to be assigned constant values in a program, this can of course be done using the LET statement as before.

```
10  A=72
20  B=63
30  A$="FLOOR AREA"
```

will allocate these values to the variables A, B and A$.

The READ statement provides an alternative method which can be useful in certain cases and is used in starter pack 6 to fill the array C( ). It could also be used in program 3.

The form of the statement is:

```
10 READ A,B,A$
```

This statement will cause the first available element in the DATA list to be read into A, the next into B and the third into A$. Of course, this means that there must be a DATA list somewhere in the program. This list may, in fact, be anywhere in the program and is ignored by the system except when a READ statement is encountered. It is good practice to keep all DATA lines together as a block of lines in a program. For the above example there must be a line in the program of the form:

```
200 DATA 72,63,"FLOOR AREA"
```

If a large amount of data is required, it may occupy more than one line of the program. Each line must begin with the DATA statement and may contain any number of data values. The number of values which can appear in a single line is only limited by the maximum line length of your machine.

As the list is read in sequence, it follows that the values in the DATA list must be in the correct order. It is also important to remember that they must be of the correct type, i.e. numeric or alphanumeric, and if of the latter type they must be contained within double quotes. If the wrong type is encountered when reading the list, an error will be signalled. There must also be sufficient data contained in the DATA list to match the READ statement otherwise this will also produce an error message.

When data is to be read into an array variable it could, of course, be done by listing the elements of the array in the correct order as follows. Suppose that in program 3 the values of C$( ) are to be allocated in this way:

```
20  READ C$(1),C$(2),C$(3)
40  DATA "CONCRETE (1:2:4)        1.44   3.04   6.40",
       "CONCRETE (1:3:6)     1.44   4.56   9.60",
       "CONCRETE (1:4:8)     1.44   6.08  12.80"
```

will solve the problem and the new lines 20 and 40 replace the lines 20, 30 and 40 in the example program.

This method is rather tedious, especially for large arrays, and an easier way would be to use the FOR-TO-NEXT statement as referred to in the previous chapter and described more fully in the next.

The lines

```
20  FOR I=1 TO 3
25  READ C$(I)
30  NEXT I
```

will produce the same result as line 20 above.

Some computers have special facilities for handling arrays called MAT functions, short for MATrix. One of these provides the capability to specify that all the values of the matrix or array are to be read. The statement

```
20 MAT READ C$
```

is equivalent to the previous lines 20 to 30 or the earlier line 20. The DATA line or lines must of course contain sufficient values to fill the array, otherwise an error will be signalled.

It has already been mentioned that if insufficient DATA appears in a list, an error message is signalled. When a DATA line or lines appear in a program, the system, when first scanning through the program, sets up a pointer at the beginning of the first DATA line. As a READ statement is processed, the pointer moves forward one place each time a value in the list is used. If the end of the list is reached before the READ statement is complete, the pointer cannot be moved forward and the error message appears.

In certain cases it may be desirable to use the values contained in a DATA statement more than once during the running of a program. To enable this to be done, the pointer must be RESTOREd to the beginning of the DATA and for this purpose the statement RESTORE is provided.

Suppose that the array N(2,3) is to be filled as follows:

Row (1)   1   2   3
Row (2)   1   2   3

```
10  DIM N(2,3)
20  FOR I=1 TO 2
30  FOR J=1 TO 3
40  READ N(I,J)
50  NEXT J
60  RESTORE
70  NEXT I


100  DATA 1,2,3
```

will achieve this result.

Manipulation of string variables

So far, the methods of assigning values to string variables have been by using the assignment statement, e.g. A$="FLOOR AREA", the INPUT statement or by READing the value from a DATA list.

Modern versions of BASIC usually include a statement or statements which make it possible to access parts of a string. The method of doing this and its flexibility will vary from one machine to another. The statements provided in the version of BASIC used by the Commodore PET and the Apple are LEFT$(, MID$(, and RIGHT$(and are not as flexibile as some, in that they can only be used to extract parts of a string and do not permit the input of data to specific portions of the string. This means that they must always appear on the right hand side of the equals sign.

```
10  B$="ABCDEFGH"
20  A$=LEFT$(B$,6)
```

will set A$ equal to the first six characters of B$ so that A$ will be set to ABCDEF. Trying to use this in the reverse fashion:

```
10  A$="ABCDEF"
20  LEFT$(B$,6)=A$
```

in an attempt to set the first six characters of B$ equal to A$, will produce an error.

A situation where this might be desirable would be in a list of the names of clients. If the list is to be sorted into alphabetical order at some time (see next chapter for the method of sorting such a list), the surnames must be the first item in the string. The surnames would be followed by the initials or the full Christian names and finally the title, Mr, Mrs etc. If the list is to be used in conjunction with other strings which contain the corresponding addresses in order to print the names and addresses for mailing, the details in the strings containing the names etc. have to be printed in reverse order — title, initials, surname.

One way of obtaining this result is to store each part of the data in a fixed field within the string variable. For example, the first sixteen characters could be reserved for the surname, the next eight for the initials and the last five for the title. Suppose

N$="SMITH   A.J.   MR   "

where the initials begin at the seventeenth character and the title at the twenty-fifth character.

If your machine has a statement which may be used on the left hand side of the equals sign, STR( ), SUB$( ) etc., information typed in using an INPUT statement can easily be allocated to its correct field within N$ in the following way.

```
20  INPUT "SURNAME",A$
30  STR(N$,1,16)=A$
40  INPUT "INITIALS",B$
50  STR(N$,17,8)=B$
60  INPUT "TITLE",C$
70  STR(N$,25,5)=C$
```

It will probably be possible to dispense with the variables A$, B$ and C$ and simply use:

```
30 INPUT "SURNAME",STR(N$,1,16)
```

etc.

As mentioned earlier in this chapter, the appearance of an empty string variable will not be the same on all machines. In some machines an 'empty' string is filled with spaces and in others it does not exist until some value is assigned to it. The method used becomes important when overcoming the inability of some microcomputers to force an INPUT into a particular field within a string as it introduces a new statement LEN( ). The object of the LEN( ) statement is to return the length of a string variable as a numeric value and is in the form:

20 X=LEN(N$)

which will set X equal to zero assuming that no value has been given to N$. If this statement is the first reference to N$ in your program, the effect of the statement with respect to N$ will depend on the machine. If an undimensioned empty string normally consists of sixteen spaces, then sixteen bytes of memory will be allocated to N$ and filled with blanks. If the string does not exist until given some value, the effect of the statement is to record the name N$ in memory but no space is allocated for its value. Whichever method your machine uses the length of the string is zero although in the first case sixteen bytes of memory have been reserved.

10 N$=" "
20 X=LEN(N$)

will also have different effects. Line 10 sets N$ to a space, so that if string variables are normally spaces when empty, the value of X is still zero as the variable N$ has not been changed. If the variable has no space allocated to it until some value is given, a space is put into memory as the value of N$ and X will be set to 1. Hence:

10 N$="SMITH   "
20 X=LEN(N$)

will set X=5 or X=16 depending on the machine assuming that there are eleven spaces after 'SMITH'.

 The following routine can be used on the Commodore PET or the Apple to input information into specific fields within a string.

10 SP$="   "
20 INPUT "SURNAME";A$
30 X=LEN(A$)
40 N$=A$+LEFT$(SP$,1,16—X)
50 INPUT "INITIALS";B$
60 X=LEN(B$)
70 N$=N$+B$+LEFT$(SP$,1,8—X)
80 INPUT "TITLE";C$
90 X=LEN(C$)
100 N$=N$+C$+LEFT$(SP$,1,5—X)

In line 10 a string variable, SP$, is set to sixteen spaces and is then used to pad out the various fields with spaces. The actual number of spaces which will be required to complete each field is calculated from the length of the input data and from the field length.

 An alternative method of doing this is to add all sixteen spaces to each of the variables A$, B$ and C$ in turn and then use LEFT$(to produce a string of the correct length. If this method is used, the length of each variable does not have to be known so that line 30 is not required. Line 40 would now be:

40 N$=LEFT$(A$+SP$,16)

Lines 70 and 100 would be modified in a similar manner.

The next problem is to arrange to print out the data in N$ in reverse order — that is, title, initials, surname.

200 PRINT RIGHT$(N$,5)+MID$(N$,17,8)+LEFT$(N$,16)

will certainly print the portions of the string in the desired order but unfortunately will include all the spaces which have been put in, or appear naturally as padding.

The method of overcoming this will once again depend on the machine. Dealing first with the type of machine which uses spaces to pad out empty strings and given that the statement used for accessing parts of a string is STR( ), the following lines will be suitable:

200 X=LEN(STR(N$,25,5))
210 PRINT STR(N$,25,X+1);
220 X=LEN(STR(N$,17,8))
230 PRINT STR(N$,17,X+1);STR(N$,1,16)

In this case line 200 finds the length of the valid information in the field containing the title and line 210 prints this field length plus one so that a space occurs between the title and the initials. Lines 220 and 230 do the same for the initials, line 230 continuing to print the surname. As this has been assumed to be the end of the PRINT statement, the trailing spaces are not important. If they were, the same procedure could be adopted as before. The four lines above could also be written as a single statement, the variable X not being used, which will dispense with lines 200 and 220. The value of X in each case is inserted in the PRINT statement as follows:

210 PRINT STR(N$,25,LEN(STR(N$,25,5))+1);STR(N$ etc.)

To do the same using a microcomputer on which spaces will be counted by the LEN( ) statement will obviously require different treatment. It will be necessary to write a routine which will act as a modified form of LEN( ) which will count the number of characters up to the first space.

200 L=5
210 C=0

```
220  FOR I=1 TO L
230  IF MID$(N$,I+25,1)=" " THEN I=L: GOTO 250
240  C=C+1
250  NEXT I
260  PRINT MID$(N$,25,C+1);
270  L=8
280  C=0
290  FOR I=1 TO L
300  IF MID$(N$,I+16,1)=" " THEN I=L:GOTO 230
310  C=C+1
320  NEXT I
330  PRINT MID$(N$,17,C+1);LEFT$(N$,1,16)
```

A loop is set up to test each character in the required field and the counter C is set to the number of valid characters — the same as the variable X in the earlier example. Note that lines 210—250 and lines 280—320 are almost identical, the only difference occurring in lines 230 and 300. They could easily be made identical by substituting another variable for the figures 25 and 16. It would not then be necessary to repeat the same lines twice as a sub-routine could be used. We shall discuss sub-routines in the next chapter.

An alternative approach to this problem is as follows:

```
200  C$=MID$(N$,25)
230  IF LEN(C$)>1 AND RIGHT$(C$,1)=" " THEN
     C$=LEFT$(C$,LEN(C$--1)):GOTO 230
260  PRINT C$;
```

In this case a string variable, C$, is set equal to the array N$ but starting at the 25th character. Line 230 then checks if the last character is a space and, if it is, it is discarded and the program returns to the beginning of line 230. If it is not, the program continues on the next line, 260, which prints the value of C$ — now with no trailing spaces. The remainder of N$ can be treated in the same way.

A practical example of splitting a string variable into a number of fields can be seen in program 3. Values are given to the arrays C$( ), M$( ) and P$( ) at the beginning of the program (lines 20 to 150) and it can be seen that spaces are left between the various pieces of information. Taking C$(1) as an example, the first portion of the string contains the description "CONCRETE (1:2:4)" which contains 16 characters. The length of field actually allocated to the description is 20 characters which allows a certain amount of latitude should any alterations be required. The string is padded out to the correct field length using spaces. This description is transferred to J$(1) in line 200 when K=1.

200 J$(K)=LEFT$(C$(K),20)

which sets J$(1)="CONCRETE (1:2:4)".

Further use of this description is in lines 340 and 360 where the proportions of coarse and fine aggregates are extracted.

340 R1$=MID$(C$(Z),13,2)

will extract the 13th and 14th characters from C$(Z). Therefore if Z=1 then R1$ is set to "2:".

350 R1=VAL(R1$)

converts this from string to number format so that it can be used in the calculations which follow. This conversion will cause no problem on the Commodore PET or the Apple as the conversion continues until a non-numeric character is encountered, in this case the colon. In other systems this is not always the case and an error will be signalled when the colon is encountered. To avoid this happening, all that is necessary is to read off the numeric value "2" only, by modifying the line to:

340 R1$=MID$(C$(Z),13,1)

Line 360 extracts the 15th and 16th characters from C$(Z) so that if Z=1 then R$ is set to "4)". The same applies here that some systems will signal an error when the bracket is encountered. To prevent this the line can be modified to:

360 R2$=MID$(C$(Z),15,1)

The next pieces of information are the quantities of cement, coarse aggregate and fine aggregate which are required for one cubic metre of concrete. Each of these is allocated a field length of 6. The quantity of cement is in the first of these fields and starts at the 21st character of the string. The first two are blank spaces and the figure 1.44 occupies spaces 23, 24, 25 and 26. This is extracted in line 255 and put into Q1$. Line 260 converts this to the numeric value Q1 for use in the calculation in line 380.

255 Q1$=MID$(C$(Z),21,6)

The field which contains the quantity of coarse aggregate starts at the 27th character and is extracted on line 265. The field containing the quantity of sand is the last 6 characters and is extracted on line 280 using this time:

280 Q3$=RIGHT$(C$(Z),6)

The same principle is used in M$( ) and P$( ). It follows, therefore, that care is needed when typing in a program line of this kind as the number of spaces between the portions of the information is very important.

Conversion of string to number format has been referred to above where VAL( ) was used in line 350. The converse of this is also possible and in Commodore PET or Apple BASIC the function is STR$( ). Although BASIC generally allows this conversion, the names of the functions are not standard.

Another use of the LEN( ) function, which was introduced earlier in this chapter, is to limit the number of characters which may be entered into a string variable. Suppose that a string variable is to have values assigned to fixed fields using the INPUT statement; it is important that the information typed in during the running of the program does not exceed the space allocated to the field.

Using the previous example, the description of the concrete in C$(1) is limited to 20 characters. If this were to be fed into the program using an INPUT statement the following routine could be used to ensure that this length is not exceeded.

```
1050  N=20
1060  INPUT A$
1070  X=LEN(A$)
1080  IF X<N+1 THEN 1120
1090  PRINT "THE MAXIMUM NUMBER OF CHARACTERS
      IS";N
1100  PRINT "TRY AGAIN"
1110  GOTO 1060
1120  C$(1)=LEFT$(A$+"                         ",20)
```

The maximum number of characters is set to 20 in line 1050. In line 1070 and 1080 the length of A$ is found and tested. If the value of X is less than the value set for the length of A$ the program continues. Otherwise a message is printed on the screen to draw attention to the error. A$ must then be re-entered.

Up to now, it has not been possible to include the double quotes in a string variable because of the meaning attached to these by the system when using (LET), INPUT and READ.

In addition to this, the comma cannot be included unless it is enclosed within double quotes, as it is otherwise regarded by the system as a terminator.

There remains one further method which can be used for assigning values to alphanumeric variables which overcomes this problem, and this is:

The GET statement

This is the name used by the Commodore PET and the Apple, but it may also appear as KEYIN or INKEY etc. on other machines. The rules for its use may vary slightly from one machine to another.

The general idea of this statement is that the system stops the execution of the program when the statement is encountered and waits until a character is ready. When a key is touched, the value indicated by that key is accepted by the system and program running continues. It is up to the programmer to decide how it is to be dealt with.

A simple use of the statement is to stop program operation, perhaps to give the operator time to read instructions which have been printed on the screen and then to continue when the operator is ready to proceed. For example,

```
90   PRINT "TOUCH ANY KEY TO CONTINUE"
100  GET X$
110  IF X$=" " THEN 100
```

This is a method of using GET which is suitable for the Commodore PET or the Apple. Processing does not actually stop, but if no key has been touched a 'null' character is assumed. This is denoted by the pair of double quotes with no space between them. In this case, as no character is available, line 110 returns the operation of the program to line 100. On some other systems, operation stops at the GET (or its equivalent) statement and waits for a value to be entered.

The GET statement can be used as a method of receiving data into a string variable from the keyboard. It has the advantage that any character, including the double quotes and the comma, can be inserted into the variable. As the characters are received one at a time, it is possible to build in a check routine so that each character is tested as it is typed in. By doing this it is possible to reject automatically any characters you wish. Suppose we have a question on the screen which requires an answer of A, B or C.

```
240  PRINT "ANSWER A, B OR C"
250  GET X$
260  IF X$<"A" OR X$>"C" THEN 250
270  REM NEXT PROGRAM LINE
```

When this program is run, lines 250 and 260 will form a loop which will continue until a letter A, B or C is typed in. If no key is touched a null character is assumed and if any key other than A, B, or C is touched it will have no effect as in either case the test in line 260 fails and the program returns to line 250. When one of the correct keys is touched, program operation will continue without the use of the RETURN key.

To use this statement to fill a string variable with characters, the following routine may be used:

```
50   REM – TO FILL THE STRING VARIABLE A$
60   A$=" "
70   GET X$
80   IF X$=" " THEN 70
90   IF X$=CHR$(13) THEN 130
100  A$=A$+X$
120  GOTO 70
130  PRINT A$
```

In order to terminate the input we type in a carriage return. The character received into X$ in line 70 will then be the code which means "carriage return" to the machine. The code for this is 13 and it is written CHR$(13). We have already made use of this system of notation when we used CHR$(147) as a method of clearing the screen when using a Commodore PET. For a more detailed discussion of these codes see chapter 10. When X$ is set equal to a carriage return, the routine is terminated and the value of A$ is printed on the screen. In the program above, as each character is typed in, the character, if it is not a carriage return, is placed in the variable A$. Nothing appears on the screen until a carriage return is received.

This can easily be modified so that each character is printed as it is typed in, by adding line

```
105 PRINT X$;
```

The semicolon is used here to keep the cursor on the same line so that the characters follow one another. There is no longer a need to print A$ at the end of the typed-in information, as it is already on the screen, so line 130 can be modified to:

```
130 PRINT
```

The line is not deleted completely as it is now used to terminate the value of A$ and bring the cursor down to the next line.

We mentioned earlier that the LEN( ) function can be used in conjunction with INPUT, to limit the number of characters which a string variable may receive. This is not necessary when using the GET statement as we can insert a counter in the program. All that is necessary is to set the value of the counter to the desired maximum.

The following revised version of the last program incorporates a counter (P), the line 105 which will print each character on the screen as it is typed in and the maximum number of characters (N), which is set to 10 in line 55. Line 92 checks the length of the string and if it is full, line 94 prints the message and line 96 terminates the INPUT as though a RETURN had been typed instead of the last character. The actual character is lost.

```
50   REM — TO FILL THE STRING VARIABLE A$
51   REM — WITH A MAX OF 10 CHARACTERS
55   N=10
60   A$=" "
65   P=1
70   GET X$
80   IF X$=" " THEN 70
90   IF X$=CHR$(13) THEN 130
92   IF P<N+1 THEN 100
94   PRINT "THE STRING IS FULL"
96   GOTO 130
100    A$=A$+X$
105    PRINT X$;
110    P=P+1
120    GOTO 70
130    PRINT
```

No provision has been made in this program to deal with the case when a mistake in entry has occurred. If this is to be included, an additional line 85 is needed to detect if the value of X$ is a backspace and, if it is, to reduce the counter P by one and to remove the last character from A$. The following extra lines could be used:

```
85   IF X$=CHR$(29) THEN 125
125    P=P—1
126    IF P=0 THEN 60
127    A$=LEFT$(A$,P)
128    PRINT X$;
129    GOTO 70
```

The statement CHR$( ) has already been used on several occasions and is explained more fully in chapter 10. It should be mentioned at this stage that it provides a further method of forcing characters such as double quotes into strings, when using the assignment statement. The code for double quotes is CHR$(34) so that:

```
200  A$=CHR$(34)+"ABC"+CHR$(34)
```

will set A$ to "ABC" with the double quotes included.

So it can be seen that there are several ways of giving values to variables in BASIC and it is up to the programmer to choose between them and use the method best suited to a particular purpose.

Most of the refinements shown in this chapter will not be met until the reader has established confidence in using simple input and assignment statements together with LEFT$( ), RIGHT$( ), MID$( ) and LEN( ). The refinements are included however because they assist in checking and program efficiency as programs become more complex.

Program 3 Estimating the price of concrete.

Program

```
10  DIM J$(3),M$(4),C$(3),P$(7)
20  C$(1)="CONCRETE (1:2:4)        1.44   3.04   6.40"

30  C$(2)="CONCRETE (1:3:6)        1.44   4.56   9.60"
40  C$(3)="CONCRETE (1:4:8)        1.44   6.08  12.80"
50  M$(1)=" 5/3.5         1.00       1"
60  M$(2)=" 7/5           1.40       2"
70  M$(3)="10/7           2.00       3"
80  M$(4)="14/7           2.80       4"
90  P$(1)="FOUNDATIONS (NE.150MM)           2.5"
100 P$(2)="FOUNDATIONS (150-300MM)          2.0"
110 P$(3)="FOUNDATIONS (EXC.300MM)          1.5"
120 P$(4)="MASS CONCRETE                    1.5"
130 P$(5)="BEDS(NE.150MM)                   5.0"
140 P$(6)="BEDS(150-300MM)                  4.0"
150 P$(7)="BEDS(EXC.300MM)                  3.25"
160 REM CALC.MATERIAL COST**************

170 PRINT"      BUILD-UP OF CONCRETE RATES"
175 PRINT

180 PRINT"SELECT THE CONCRETE MIX YOU REQUIRE:"
185 PRINT
190 FOR K=1 TO 3
200 J$(K)=LEFT$(C$(K),20)
210 PRINT J$(K);
220 PRINT"          (";K
230 NEXT K
240 INPUT"THE NUMBER OF MY SELECTION IS ";Z
250 PRINT CHR$(147)
255 Q1$=MID$(C$(Z),21,6)

260 Q1=VAL(Q1$)

265 Q2$=MID$(C$(Z),27,6)
270 Q2=VAL(Q2$)
280 Q3$=RIGHT$(C$(Z),6)
```

Operation

10 Dimensions the array variables.

20—150 Assigns values to the variables in the arrays.

20—40 Table shows: concrete mix; volumes of cement, sand and aggregate.

50—60 Table shows: mixer size; output in m^3 per hour; selection number.

90—150 Table shows: placing position of concrete; labour hours required to place one m^3 in this position.

160 Acts as a signpost. Has no effect on program.

170 Prints title.

175 Leaves a blank line on the screen under title. Also used on other lines in this program.

180 Prints heading.

190—230 Prints list of choices.

240 Asks for choice of concrete mix.

250 Clears screen.

255 Sets Q1$=quantity of cement per cubic metre for chosen mix.

260 Converts the string to a numeric value Q1.

265—290 Does the same for coarse aggregate & sand.

Program 3 Contd.

Program

```
290 Q3=VAL(Q3$)
300 PRINT"ENTER CURRENT COST OF MATERIALS:"
305 PRINT
310 INPUT"   CEMENT      ";CEM
315 PRINT
320 INPUT"   SAND        ";SAN
325 PRINT
330 INPUT"   AGGREGATE   ";AGG
340 R1$=MID$(C$(Z),13,2)
350 R1=VAL(R1$)

360 R2$=MID$(C$(Z),15,2)
370 R2=VAL(R2$)
380 CM=(((Q1*CEM)+(Q2*SAN)+(Q3*AGG))*1.25)/(1+R1+R2)

400 REM CALC. MIXING COST(MACHINE)******
410 PRINT CHR$(147)
420 PRINT"SELECT YOUR MACHINE FOR MIXING:"
430 PRINT
440 PRINT"SIZE(CF)     OUTPUT/M3     NO.MEN"
450 PRINT
460 FOR K=1 TO 4
470 PRINT M$(K);
480 PRINT"         (";K
485 PRINT
490 NEXT K
500 PRINT"THE NUMBER OF MY SELECTION IS ";Y
510 PRINT CHR$(147)
520 PRINT"YOUR CHOICE IS AS FOLLOWS:"
530 PRINT
540 PRINT M$(Y)
550 PRINT
560 INPUT"HOURLY COST FOR THIS MACHINE IS:-";HC
565 PRINT
570 INPUT"% ADDITION FOR FUEL,MAINT,ETC. :-";P
575 PRINT
580 INPUT"HOURLY COST FOR LABOURERS     :-";LC
590 HC=HC+(HC*P/100)
```

Operation

300 Prints heading.

310—330 Asks for costs of materials to be INPUT.

340 Sets R1$=proportion of coarse aggregate.
350 Converts the string to a numeric value R1.
360 & 370 Do the same for sand.

380 Sets CM to cost of materials to make 1m³ of concrete.
400 Acts as signpost — has no effect on program.
420 & 440 Print headings.

460—490 Print list of choices for mixer.

500 Asks for choice of mixer.

520 & 540 Print chosen mixer size.

560—580 Asks for mixer costs.

590 Sets HC to hourly cost plus % for fuel etc.

Program 3 Contd.

Program

```
600 MX$=MID$(M$(Y),15,6)
510 MX=VAL(MX$)

620 MY$=RIGHT$(M$(Y),2)
630 MY=VAL(MY$)
640 M=(HC+(MY*LC))/MX

550 BC=M+CM
660 PRINT
670 PRINT"          COST SUMMARY"
680 PRINT
690 PRINT"PLAIN CONCRETE IN:-        COST/M3"
700 PRINT
710 FOR K= 1 TO 7
720 D$=LEFT$(P$(K),24)
730 L$=RIGHT$(P$(K),6)
740 L=VAL(L$)
750 T1=BC+(L*LC)

752 T1=(T1*100)+.5
753 T1=INT(T1)
754 T1=T1/100
760 PRINT D$,T1
770 PRINT
780 NEXT K
800 END
```

Operation

600 Sets MX$=quantity of concrete/hour.
610 Converts the string to a numeric value.
620—630 Does the same for number of men required.
640 Calculates labour cost per cubic metre.
650 Calculates total cost per cubic metre.

670 & 690 Prints headings.

710—780 Prints cost summary.
720 Sets D$ to description.
730 Sets L$ to the labour cost factor.
740 Converts this to a numeric value L.
750 Adds the labour cost factor for the particular situation.
752—754 Rounds the value to two decimal places.

760 Prints the description and cost.

800 Stops the program.

CHAPTER 8

Choosing and sorting

Choosing

In the office of the design team many situations arise where much time may be spent in searching for particular items of information — for example, in using the Barbour (or similar index) to find the names and addresses of the manufacturers of particular building components. If this information is held in a suitable data base, a program can be written to access this data and rapidly extract the desired information after the requisite questions appearing on the screen have been answered. A data base such as this will be quite large and will therefore take some time to compile and, of course, the machine must have sufficient storage capacity in the backing stores to contain it.

In the previous chapter we have discussed the various methods which may be used to supply a computer with data and the data base can be built up using these techniques. For this information to be of use, it must be readily accessible and the right information needed for a particular situation must be available as quickly as possible and in a form in which it can be understood. This chapter deals with the methods which can be used to achieve this result.

A simple example of this type of operation is given in program 4(a) (at the end of the chapter), which is a program containing the kind of information which might be held in the office files of an architect. When a new job is received, it is useful to be able to check back on the records to see whether any similar building has been constructed recently (see fig. 8.1).

By storing this information in the computer, the necessary references can easily and quickly be found. For simplicity the information stored in this example is limited to the following:

1. Office file reference
2. Type of building
3. Floor area
4. Number of storeys.

This can be increased to include any other information which may be desired.

In this program the data is supplied in the form of READ statements

Fig. 8.1 Flow chart for program 4(a): search records 1.

and DATA lines which, of course, limits its usefulness. With the amount of data which has been included in this way it is obvious that no great feat of memory would be required to recall the particular jobs concerned. The next chapter will deal with the use of computer files and this program will be developed further at that stage which will greatly enhance its range.

By answering a series of questions which are programmed to appear on the screen, the required parameters are found and the data is searched for references which suit the parameters.

For the purposes of this program, each data entry is provided with a simple code.

The code is:

Type of building: 1. School
                     2. Factory
                     3. Detached house
                     4. Semi-detached house
                     5. Terraced house
                     6. Flats

Floor area: 1. 0—99 m²
                     2. 100—199 m²
                     3. 200—299 m²
                     4. 300—399 m²
                     5. 400—499 m²
                     6. over 500 m²

Number of storeys: 1—6

Therefore a factory with a floor area of 450 m² and two storeys would be coded 252.

The answers to the questions which are programmed to appear on the screen are placed in the string variable A$ in lines 360—380. A$ is then compared with the codes in the data bank contained in lines 700—730.

The operation of building up the string variable A$ using the numeric answers to the questions involves the use of a new statement, STR$( ). This statement is used to convert a numeric value into a string value, so that:

20 N=5
30 A$=STR$(N)

will first set the numeric variable N equal to 5 in line 20. Line 30 then converts this into the string variable A$ which is also set equal to 5.

The reverse procedure is accomplished by using VAL( ) in the

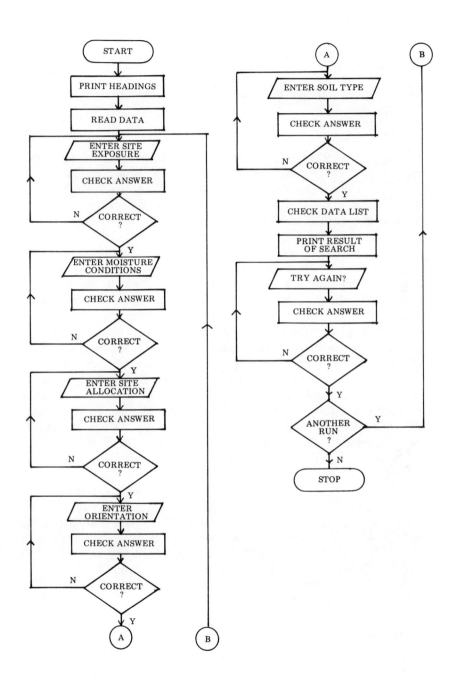

Fig. 8.2 Flow chart for program 5: trees.

following way:

20 A$=5
30 N=VAL(A$)

These two statements therefore make possible the conversion of the two variable types.

A different problem arises in program 5 (at the end of the chapter), which is designed to select suitable species of trees and shrubs for a particular requirement, based on exposure, moisture conditions, site location, orientation and soil type (see fig. 8.2). This program is based on a routine written by Julia Hunt when a student at the Welsh School of Architecture, UWIST, Cardiff.

In this case many of the trees and shrubs are suitable for more than one of each of the specified conditions and so cannot be assigned a simple reference code as was the case in program 4(a).

The method used in program 5 is to store the suitability of the various species in a two-dimensional array which can be seen in lines 4120—4240. The rows represent the various types of tree or shrub and the columns represent the suitability. A figure one indicates that the tree or shrub is suitable and zero that it is not.

Five degrees of exposure are possible in the program, so that the first five columns refer to exposure. The next question relates to moisture conditions, so that columns 6—8 refer to moisture etc. A total of 19 columns is used although the array is dimensioned to accept a maximum of 20 columns. Also the array can take up to 30 varieties of trees or shrubs, although only 25 are at present in the data bank. This number is set in line 110. If it is desired to increase the scope of the program, the dimensions of the arrays in line 90 can be adjusted to suit.

When all the questions have been answered, the program uses a FOR-TO-NEXT loop (lines 820—1530) to check through the list in order to discover if the required elements in the array A( ), which have been referenced by the answers to the questions, are set to 1. If all five satisfy this condition, the name of the tree or shrub is printed. If no suitable species is found, this is stated.

Some more BASIC statements will be dealt with in more detail, or for the first time, in this chapter.

FOR-TO-NEXT statements
These statements have already been used in the two previous chapters and their purpose is to set up a loop. All the statements within the loop are performed a prescribed number of times. A simple example of this is contained in program 4(a).

```
30  FOR I=1 TO V
40  READ K$(I)
50  NEXT I
```

This is similar to the way in which it was used earlier, to read values into an array from a DATA list, but this time the length of the loop will depend on the value given to V. The value of V in this program is set in line 26. If further data is added to the program, it is only necessary to change the value of V in this line for the whole of the data to be read.

It is also possible to use a variable for the starting value of the loop, e.g.

```
30  FOR I=X TO V
```

where X and V have been given values in an earlier line.

The purpose of the line containing the FOR-TO statement is to set up the parameters of the loop which are placed in the memory when the line is processed. When the line containing NEXT is reached, the value of the variable is changed and program operation returns to the statement following the FOR-TO statement unless the preset limit has been reached, when the parameters are cleared from memory and the program continues with the statement which follows the NEXT instruction.

The examples in the previous chapter all used the default value of +1 as the amount to be added to the variable each time the NEXT instruction was encountered. If some other value is required as was the case with program 1, this is given as the STEP, e.g.

```
30  FOR I=10 TO 1 STEP −1
```

will cause the value of I to be reduced by one each time the NEXT statement is encountered. Note that:

```
30  FOR I=10 to 1
```

or

```
30  FOR I=1 to 10 STEP −1
```

will cause an error on some machines because the end of the loop could never be reached. On other machines the loop will be processed once only.

In most of the modern versions of BASIC, termination of the loop occurs when the value of the variable has either reached or exceeded the preset limit.

```
30  FOR I=1 TO 9 STEP 3
```

would take values of I as 1, 4 and 7; the next value, 10, being too great, would cause the loop to terminate.

FOR-TO-NEXT loops may be 'nested', i.e. one inside the other, and there may be no limit to the number of nestings which may take place. There may be restrictions on some microcomputers depending on the number of variables used and if the loops are left before the NEXT statement is reached.

```
10  DIM A$(3,5)
20  FOR I=1 TO 3
30  FOR J=1 TO 5
40  READ A$(I,J)
50  NEXT J
60  NEXT I
```

enables the DATA for A$( ) to be read — a suitable DATA line must of course appear somewhere in the program. It is usually possible to combine NEXT statements when a number of loops terminate at the same point in the program, so that lines 50 and 60 above could be combined into

```
50 NEXT J, I
```

It is, of course, important to ensure that the loops are terminated in the correct order — the innermost loop first etc.

A FOR-TO statement without a corresponding NEXT statement will not cause an error. As already explained, its only function is to set the parameters. A NEXT statement without the corresponding FOR-TO statement *will* cause an error because this attempts to access the parameters which should have been set, and when it does not find them an error is signalled. It follows that it is possible to leave a loop before it is completed, but it is not possible to jump into a loop bypassing the FOR-TO statement.

Although it is possible to leave a loop without going through the NEXT statement, this will mean that the parameters remain in memory, so that if this process is repeated a sufficient number of times the available memory space will be filled with unwanted parameters and an error will be signalled when no more memory space is available.

It is therefore not good practice to leave a loop this way. Suppose the array A(100) may contain an indefinite number of values which have been assigned to it in lines 20—90 and we wish to find out how many values there are.

```
10  DIM A(100)
```

```
100  FOR I=1 TO 100
110  IF A(I)=0 THEN 130
120  NEXT I
130  PRINT "THE NEXT FREE SPACE IS A(";I;")"
```

will set I to the first element of A( ) which equals zero. In doing so the program leaves the loop and program operation continues, leaving the parameters of that loop in memory.

There are various ways in which this can be avoided, one of which would be:

```
10   DIM A(100)

100  FOR I=1 TO 100
105  X=I
110  IF A(I)=0 THEN I=100
120  NEXT I
130  PRINT "THE NEXT FREE SPACE IS A(";X;")"
```

This time the additional variable X is set to the current value of I each time the loop is processed.

When line 110 tests for the value of A(I), instead of branching to line 130, I is set to the upper limit so that when NEXT is encountered in line 120 the loop is terminated and the parameters are cleared from memory. The variable X contains the last value of I to be used and hence is the element number of A( ) which satisfies the conditional statement in line 110. No provision has been made in this example for the case when the array A( ) is full. To take into account this possibility, the program can be modified as follows.

```
10   DIM A(100)

95   D=0
100  FOR I=1 TO 100
105  X=I
110  IF A(I)=0 THEN I=100:D=1
120  NEXT I
125  IF D=0 THEN PRINT "ARRAY FULL":GOTO 140
130  PRINT "THE NEXT FREE SPACE IS A(";X;")"
140  STOP
```

In this version another variable D has been introduced. This is set to zero before the FOR-TO-NEXT loop is entered. You will notice that line 110 contains two separate statements:

(i) IF A(I)=0 THEN I=100
(ii) D=1

These two statements are separated by a colon and this is a method commonly used to save space in memory and to speed the operation of a program. This is discussed further in chapter 10. In the programs included within the chapters of this book we have, in general, written them with one statement to each line for clarity. In this case, however, it is necessary to put the two statements together on a single line as this has a special meaning to many of the microcomputers. The machine will interpret this to mean that if the conditional statement is true, process the rest of the line, otherwise go to the next statement line. Therefore D is set to 1 in line 110 if and only if a zero is found in the array. The same thing occurs in line 125 which detects if D=0 and, if it is, prints the 'ARRAY FULL' message; otherwise the original line 130 prints the message giving the number of the next free element.

If you find that this does not work on your machine the following version of the program, though somewhat longer, should be satisfactory:

```
10    DIM A(100)

95    D=0
100   FOR I=1 TO 100
105   X=1
110   IF A(I)=0 THEN 115
112   GOTO 120
115   I=100
116   D=1
120   NEXT I
125   IF D=0 THEN 135
130   PRINT "THE NEXT FREE SPACE IS A(";X;")"
131   GOTO 140
135   PRINT "ARRAY FULL"
140   STOP
```

IF-THEN statement

The previous example has introduced the conditional statement in the form IF-THEN, which allows decision making during the run of a program. The relationship tested in that example was to check when the value of A(I) was equal to zero. The other relational operators may also be used, i.e. $<$, $>$, $<=$, $>=$ and $<$ $>$.

In its simplest form, if the relationship is true, the program transfers to the line number specified, i.e.

```
110 IF A(I)=0 THEN 130
```

or

110 IF A(I)=0 GOTO 130

or

110 IF A(I)=0 THEN GOTO 130

The form will vary from one machine to another. Some micro-computers will accept all three versions.

Modern versions of BASIC allow greater freedom with this statement in that a number of relationships can be tested in the same statement using the logical operators AND and OR. If the relationship is satisfied the statement or statements following are executed, otherwise the program continues on the next line.

150 IF A>B AND A<C AND B<>0 THEN A=B*C

means that if *all* the relationships are true then A=B*C

150 IF A$=T$ OR B$=C$ THEN STOP

means that if *either* of the relationships is true then the program stops.

GOSUB-RETURN statements

Program 4(a) has introduced the concept of the sub-routine. This is a useful device which can be used to avoid repeating portions of a program.

It is good practice when writing a program to ensure, as far as possible, that wrong entries cannot be made. In Program 4(a) the various questions which appear on the screen are arranged so that each question requires a numerical answer. In each case, the lowest possible valid number is 1 while the upper limit varies from one question to another. Following each question, the variable D is set to the upper limit for that particular question and the program branches to the error testing sub-routine in lines 1000—1020. This sub-routine checks that the number typed in lies between 1 and the given upper limit in the variable D and that it is an integer. If it fails these tests the message "ERROR IN ENTRY — TRY AGAIN" is printed on the screen and the variable D is set to 1. If the test is passed, the variable D is set to zero. The value of D is tested on return from the sub-routine; if D=1 the program returns to ask the question again, and if D=0 the program continues. Note that the variable D is used in two different ways here — first to set the maximum value and then to carry the error message.

All sub-routines must finish with a RETURN statement which causes the program to resume operation at the statement following GOSUB.

A GOSUB statement sets a marker in memory which allows the program operation to RETURN to the correct point. Repeated jumping out of sub-routines will build up unwanted parameters in the memory and will therefore have the same effect as jumping out of a FOR-TO-NEXT loop. In this case it is the RETURN statement which clears the parameters from memory.

The first choice in this program is the type of building and the first selection is school, factory or domestic. Domestic is further broken down into detached, semi-detached and terraced houses, and flats.

This second part of the selection is also done using a sub-routine which in turn requires a numerical answer to a question. Once more the sub-routine in lines 1000—1020 is used to check the entry. This indicates that sub-routines can also be nested in that one sub-routine can call another which in turn can call another etc. It is common practice to place all sub-routines at the end of the program.

TAB function for format control
Earlier programs have used spaces within the double quotes to cause text to be printed at some distance from the left hand edge of the screen:

110 PRINT"   1) SCHOOL"

will leave three spaces at the left hand edge of the screen before printing the text.

110 PRINT TAB(3);"SCHOOL"

will produce the same result. The TAB function is similar to the tab stop on a typewriter and using TAB(3) means that printing is to begin at position number 3 on the screen. It must be remembered that the position numbers on the screen start at zero so that there are three spaces before position number 3. If a numeric value follows the TAB function, a space will be left for the sign of the number if the number is positive.

2000 PRINT TAB(3);N

will cause the value of the variable N to be printed beginning after three spaces with a minus sign if the value of N is negative. If N is positive the number begins after four spaces, one space having been left for the inferred positive sign.

The following routine uses the TAB function to position a text string in the centre of the screen.

10 REM SCREEN WIDTH=W
20 W=40

```
30 INPUT"ENTER TEXT STRING";T$
40 L=LEN(T$)
50 IF L>W THEN PRINT "STRING TOO LONG":GOTO30
60 PRINT TAB((W-L)/2);T$
70 GOTO 30
```

The length of the string which has been INPUT is tested in line 40 to see whether it exceeds the screen width. If it does not, the distance from the left to allow the string to be centred is calculated within the brackets of the TAB function.

When using the TAB function with an expression for calculating the position to print and the expression within the brackets does not produce a whole number, only the integer portion of the result is used. If the expression produces a negative result or a number which is less than the current cursor position, the cursor does not move. (Note: a negative result will produce an error on some machines.)

GOTO statement
This is a very easily understood statement and is used to cause a branch in program operation. It is used in the previous program to return program operation to line 30 so that different text lengths can be tried. This program will continue to loop until the operator stops it with the STOP key.

Computed branches
GOTO and GOSUB statements which direct a program to a specific line have already been dealt with. The ON statement permits the programmer to choose one of several line numbers depending on circumstances.

Program 4(a) gives the option in lines 490—500 to repeat the program (enter 1) or to terminate it (enter 2). The value of Q entered in line 500 must therefore be 1 or 2. The next line:

510 ON Q GOTO 540,560

directs that if Q=1 GOTO 540 which clears the screen and then goes to the beginning of the program. If Q=2 GOTO 560 which is the end of the program.

If the value of Q is neither 1 nor 2, no branch occurs and the program continues on the next line.

In order to use the same error sub-routine, the value of N is set to zero in line 515 so that the test will fail and the error message will be printed. No use is made in this case of the variable D which is set by the sub-routine.

It is also possible to use the ON statement in exactly the same way with GOSUB.

An expression can be used in place of Q, e.g.

60 ON T*21 GOTO 100,200,300,400

If T=.1 the result of the expression is 2.1. The machine uses the integer portion of the result and the program will branch to line 200.

Sorting

It is often convenient, or even necessary to present results in numerical or alphabetical order.

Program 4(a) presents the results of the search as a list of office file numbers. Because the DATA in lines 700—730 has been entered with the file numbers in order, the list produced will be in ascending numerical order. This is not always the case and a simple routine can be used to arrange a series of numbers into an ascending (or descending) order.

```
10  DIM A(5)
20  FOR I=1 TO 5
30  INPUT A(I)
40  NEXT I
```

allows five numbers to be assigned to the array A( ) and we can enter the numbers in any order. To arrange these numbers in numerical order a program must be written to compare the numbers in pairs, and if the two numbers compared are not in the correct order, to interchange them. This process must be repeated until the numbers have been placed in the correct order. The worst case will be when the smallest or the largest number is at the wrong end of the list.

Suppose we enter the numbers as 5, 4, 3, 2, 1 and we require the numbers to be arranged in ascending order; if the program compares the first pair, it can be seen that they are in the wrong order so that they must be interchanged. If this is done the result will be 4, 5, 3, 2, 1. The next comparison is between the second and third numbers 5 and 3 which also need to be changed round. The result of comparing each pair in turn is as follows:

5, 4, 3, 2, 1
4, 5, 3, 2, 1
4, 3, 5, 2, 1
4, 3, 2, 5, 1
4, 3, 2, 1, 5

So that after all four pairs have been compared and adjusted, the figure 5 has been moved to its correct place at the end. The comparison is now repeated for the first three pairs only, resulting in:

4, 3, 2, 1, 5
3, 4, 2, 1, 5
3, 2, 4, 1, 5
3, 2, 1, 4, 5

The last two figures are now in their correct position and can be left out of the next comparison which produces:

3, 2, 1, 4, 5
2, 3, 1, 4, 5
2, 1, 3, 4, 5

The final pass will reverse the first two figures and the list will have been placed in the correct order.

A flow chart of a program which will achieve this result was given in fig. 5.3. The actual program uses two nested FOR-TO-NEXT loops as follows:

```
50    FOR I=4 TO 1 STEP −1
60    FOR J=1 TO I
70    IF A(J)<=A(J+1) THEN 110
80    T=A(J)
90    A(J)=A(J+1)
100    A(J+1)=T
110    NEXT J,I
```

Line 70 tests to see whether an interchange is required, and if it is the interchange is done by introducing a numeric variable T which holds the value of one of the numbers during the changeover. If no change is required the next value of J is tested. To see the result of this program we can add:

```
120 FOR I=1 TO 5
130 PRINT A(I)
140 NEXT I
```

which will print out the numbers we fed in earlier, but now in ascending order.

This is only one solution to the problem; the pairs of numbers can be taken in different ways. For example, the first number can be compared with each of the others in turn and interchanged if necessary. One pass will therefore move the lowest number to its correct place at the beginning of the list. The alternative lines to do this are:

```
50    FOR I=1 TO 4
60    FOR J=I+1 TO 5
70    IF A(I)<=A(J) THEN 110
80    T=A(I)
90    A(I)=A(J)
100   A(J)=T
110   NEXT J,I
```

Alphanumeric arrays can be sorted in exactly the same way, although in this case it is advisable to be aware of the relative values of the characters which are being compared (see chapter 10 for a fuller explanation of these values). The letters of the alphabet are ordered as expected, i.e. A, B, C etc. increase in value. The numbers too, when used in a string variable, increase in the expected order. If your machine has the capability of lower case letters these will probably have a higher value than the capitals, so that 'a' is greater than 'Z'. Below the numbers in value and between the numbers and the letters are the various symbols. The lowest value is the space. For a full list of these characters and symbols and their relative values on your machine, see your instruction manual.

Suppose two strings A$ and B$ are being compared; A$="SMITH" and B$="BROWN". Comparison will stop after the first character because 'B' is less than 'S'. If A$="SMITH,A" and B$="SMITH,B" comparison will continue to the seventh character where the first difference occurs and A$ is found to be less than B$. Now suppose that B$="SMITH, B"; the seventh character of A$ is 'A' and the corresponding character of B$ is a space which has a lower value than 'A'. Hence B$ is less than A$ and the names would appear in the incorrect order.

Another point to remember is that if an array is not completely full, the elements not used will be zeros if it is a numeric array and nulls or spaces if it is an alphanumeric array. This means that unused elements will appear at the beginning of the list after sorting into ascending order or at the end of the list if sorted into descending order.

```
10  DIM N$(10)
20  PRINT "TYPE IN NAMES — TYPE '/' TO FINISH"
30  FOR I=1 TO 10
40  INPUT N$(I)
50  IF N$(I)="/" THEN I=10
60  NEXT I
```

This program will allow up to a maximum of ten names to be entered and could then be followed by one of the sort routines previously described. If only five names are entered, these will be in the array elements N$(6) to N$(10) after sorting.

To prevent this, the blank values must be changed to something which is greater than, or equal to, any that can be typed in. The highest value is CHR$(255) so the program could be modified to:

```
10  DIM N$(10)
12  FOR I=1 TO 10
13  N$(I)=CHR$(255)
14  NEXT I
20  PRINT "TYPE IN NAMES — TYPE '/' TO FINISH"
30  FOR I=1 TO 10
40  INPUT N$
50  IF N$="/" THEN I=10:GOTO 60
55  N$(I)=N$
60  NEXT I
```

This time a FOR-TO-NEXT loop has been included to preset all the values in the array N$( ) to CHR$(255). Each INPUT is into the string variable N$ and not directly into the array element; it is put into the correct element in line 55. If '/' is entered, the array element is not affected.

So far, the data which has been used in a program has either been written into the program as DATA or typed in by the operator. This will, of course, limit the amount of data which can be handled to such an extent that in many cases it would be easier to just look at the data and pick out the required information without the use of any machine. These simple examples have been used to illustrate the principles involved in dealing with data. The same principles can be applied when the quantity of data is large as would be the case in any practical application. In such a situation, when the amount of data must be very large, it is stored in files on disk or tape and pulled in by the program as it is required. The method of doing this is dealt with in the next chapter.

A computer is always limited by the amount of usable memory available for program and data storage. Large arrays of data can quite easily fill up all the available space and leave insufficient room for the program.

At the beginning of this chapter we suggested that a possible use for a data base would be to contain selected information provided by the Barbour Index or other similar system. Even if the data held is limited to the names and addresses of the manufacturers and their associated products, it would be beyond the capacity of most microcomputers to hold all this data in memory at one time. Even if the data is typed in in alphabetical order of manufacturers when the system is first set up, any future additions will upset this order so that a sort routine will be required to restore the list to a correct alphabetical sequence. We may also wish to be able to print out a list of products in alphabetical order or in product groups etc. In order to do this, several completely different arrangements of the data will be required.

A method of achieving this result is to use a number of 'locator arrays' to contain the information as to the sequence in which the data must be taken to be in the desired order.

To produce a locator array, we must first read the whole file and extract a suitable 'sort key'. If the problem is to find the sequence for an alphabetical list, the sort key will consist of the names of the various manufacturers. Suppose the file contains 500 such names we can dimension a suitable array, say N$( ), to contain them, and also a similar sized numeric array to contain the locator.

```
10  DIM N$(500),L(500)
```

dimensions the two arrays.

```
20  FOR I=1 TO 500
30  L(I)=I
40  NEXT I
```

sets the elements of the array L( ) to the numbers 1—500. The program lines which follow contain the means of allocating the names to the array N$( ) ready for sorting and the following modified version of a sort routine:

```
200  FOR I=1 TO 499
210  FOR J=I+1 TO 500
220  IF N$(L(I))<=N$(L(J)) THEN 260
230  T=L(I)
240  L(I)=L(J)
250  L(J)=T
260  NEXT J,I
```

compares the values of N$( ) in line 220 as before, but then alters the values of L( ).

```
270  FOR I=1 TO 500
280  PRINT N$(L(I))
290  NEXT I
```

prints an alphabetical list of the manufacturers. The corresponding addresses and products will be contained in the corresponding element of the relevant arrays. Using this method of sorting, no data is physically moved within the file or from one element to another in the array.

The system can be used for any collection of information that needs structuring, such as estate agents' lists, cost analyses, tables of U-values, clients' Christmas card lists etc.

The next chapter deals with the methods which may be used for handling such large amounts of data.

Program 4(a) Search records 1.

Operation

10 Clears screen.
20 Dimension arrays.
25 V set to suit the data available.
30 Set up loop to read data.

60 Leaves blank line on screen; used throughout program.

70 & 80 Print heading.

100—130 sets up first question.

135 Sets variable used in question to zero.
140 Answer question.
150 Sets error variable to maximum value.
160 To sub-routine to check answer.
165 Return to ask question again if incorrect.

180 To sub-routine if domestic.
190—250 Second question.

255 Sets variable used in question to zero.
260 Answer question.
270 Sets error variable to maximum value.
280 To sub-routine to check answer.
285 Return to ask question again if incorrect.

300 Third question.

Program

```
10 PRINT CHR$(147)
20 DIM K$(100),K(100)
25 V=12
30 FOR I=1 TO V
40 READ K$(I)
50 NEXT I
60 PRINT

70 PRINT "THIS PROGRAM WILL SEARCH RECORDS"
80 PRINT "TO FIND SIMILAR TYPES OF JOB"
90 PRINT
100 PRINT "TYPE OF BUILDING"
110 PRINT TAB(3);"1) SCHOOL"
120 PRINT TAB(3);"2) FACTORY"
130 PRINT TAB(3);"3) DOMESTIC"
135 N=0
140 INPUT "(1,2 OR 3)";N
150 D=3
160 GOSUB 1000
165 IF D=1 THEN 135

170 NO=N
180 IF NO=3 THEN GOSUB 1500
190 PRINT CHR$(147);"FLOOR AREA"
195 PRINT
200 PRINT TAB(3);"1)    0-99     SQ.M"
210 PRINT TAB(3);"2)  100-199   SQ.M"
220 PRINT TAB(3);"3)  200-299   SQ.M"
230 PRINT TAB(3);"4)  300-399   SQ.M"
240 PRINT TAB(3);"5)  400-499   SQ.M"
250 PRINT TAB(3);"6)  500+      SQ.M"
255 N=0
260 INPUT "(1,2,3,4,5 OR 6)";N
270 D=6
280 GOSUB 1000
285 IF D=1 THEN 255

290 N1=N
300 PRINT CHR$(147);"NUMBER OF STORIES"
```

Program 4(a) Contd.

Program

```
310 PRINT
315 N=0
320 INPUT "(1,2,3,4,5 OR 6)";N
330 D=6
340 GOSUB 1000
350 IF D=1 THEN 315

355 N2=N
360 A$=MID$(STR$(N),2,1)
370 A$=A$+MID$(STR$(N1),2,1)
380 A$=A$+MID$(STR$(N2),2,1)
390 FOR I=1 TO V
400 IF A$=RIGHT$(K$(I),3) THEN K(I)=1
410 NEXT I
420 PRINT CHR$(147);"RESULT OF SEARCH"
430 PRINT
435 D=0

440 FOR I=1 TO V
450 IF K(I)=1 THEN D=1:PRINT "FILE NUMBER";
    LEFT$(K$(I),8)
460 K(I)=0
465 NEXT I
470 IF D=0 THEN PRINT "NO SIMILAR JOBS FOUND"

490 PRINT "DO YOU WISH TO SEARCH FOR MORE"
495 Q=0
500 INPUT "1) YES, 2) NO";Q
510 ON Q GOTO 540,560
515 N=0
520 GOSUB 1000

530 GOTO 500
540 PRINT CHR$(147)
550 GOTO 100
560 END
700 DATA "1234    131","1235    132","1236    412"
710 DATA "1237    221","1238    312","1239    241"
720 DATA "1240    522","1241    312","1242    322"
730 DATA "1243    141","1244    665","1245    654"
```

Operation

315 Sets variable used in answer to zero.
320 Answer question.
330 Sets error variable to maximum value.
340 To sub-routine to check answer.
350 Return to ask question again if incorrect.

360—380 Set A$ to suit answers to above questions.

390—410 Search loop.

420 Prints heading.

435 Sets D to zero — the value which means 'no similar jobs'.
440—465 Loop to print results.
450 Prints file number if K(I)=1.

460 Resets K(I) to zero.

470 If D is still zero there are no similar jobs on file.
490 Prints question.
495 Sets variable in answer to zero.
500 Answer question.

520 To error sub-routine if answer incorrect.
500 Return to ask question again.
540 Clear screen.
550 Return to start of program.
560 Program run stops.
700—730 Data.

Program 4(a) Contd.

Program

```
1000 IF N<1 OR N>D OR N<>INT(N) THEN PRINT "ERROR
     IN ENTRY-TRY AGAIN":D=1:GOTO 1020
1010 D=0
1020 RETURN
1500 PRINT CHR$(147);"TYPE OF DOMESTIC BUILDING"
1505 PRINT
1510 PRINT TAB(3);"1) DETACHED HOUSE"
1520 PRINT TAB(3);"2) SEMI-DETACHED HOUSE"
1530 PRINT TAB(3);"3) TERRACED HOUSE"
1540 PRINT TAB(3);"4) FLATS"
1545 N=0
1550 INPUT "(1,2,3 OR 4)";N
1560 D=4
1570 GOSUB 1000
1580 IF D=1 THEN 1545

1590 N0=N0+N

1600 RETURN
```

Operation

1000 Error sub-routine.

1500 Sub-routine to choose type of domestic building.

1545 Set variable used in question to zero.
1550 Answer question.
1560 Sets error variable to maximum value.
1570 To sub-routine to check answer.
1580 Return to ask question again if incorrect.
1590 Revises value of N to suit answer to last question.

Program 5 Trees.

Program

```
10  PRINT  CHR$(147);TAB(8);"TREE SELECTION PROGRAM"
20  PRINT
30  PRINT
40  PRINT  "THIS PROGRAM PROVIDES A LIST OF TREES"
41  PRINT
50  PRINT  "& SHRUBS WHICH WILL GROW UNDER CERTAIN"
51  PRINT
60  PRINT  "SITE CONDITIONS:ANSWER THE FOLLOWING"
61  PRINT
70  PRINT  "FIVE QUESTIONS ABOUT YOUR SITE BY"
71  PRINT
80  PRINT  "SELECTING FROM THE ALTERNATIVES OFFERED:"
81  PRINT  "TYPE ONLY THE NO. OF YOUR CHOSEN ANSWER:"
90  DIM A(30,20),N$(30)
100 REM  T=NUMBER OF TREES IN DATA BANK
101 T=25
110 GOSUB 4000
120 PRINT  "TOUCH ANY KEY TO CONTINUE"
130 GET X$
140 IF X$="" THEN 130
310 PRINT  CHR$(147);"Q1. :SITE EXPOSURE."
311 PRINT
320 PRINT  TAB(3);"1. VERY EXPOSED."
330 PRINT  TAB(3);"2. EXPOSED."
340 PRINT  TAB(3);"3. AVERAGE."
350 PRINT  TAB(3);"4. SHELTERED."
370 PRINT  TAB(3);"5. VERY SHELTERED."
390 PRINT
400 INPUT  "(1,2,3,4 OR 5)";E
410 IF E<5 AND E>1 GOTO 420

415 GOSUB 3000
416 GOTO 400
420 PRINT  CHR$(147);"Q2. :MOISTURE CONDITIONS ON SITE."
421 PRINT
430 PRINT  "1. DAMP."
440 PRINT  "2. AVERAGE."
450 PRINT  "3. DRY."
460 PRINT
```

Operation

10 Clears screen & prints title.
20 Leaves blank line. Used throughout program.
40—81 Prints headings.

90 Dimension arrays.
100 Acts as a signpost. Has no effect on program.
110 Transfers operation to sub-routine.
120 Stops program operation to allow time to read headings.

310—370 Prints first question.

400 Answer question.
410 Check answer. Skips next two lines if correct.
415 To error sub-routine if incorrect
416 Returns to ask question again.
420—450 Prints second question.

Program 5 Contd.

Program

```
470 INPUT "(1,2 OR 3)";M
480 IF M<3 AND M>1 GOTO 520

490 GOSUB 3000
500 GOTO470
520 PRINT CHR$(147);"Q3. :URBAN POLLUTION/SITE LOCATION"
521 PRINT
530 PRINT TAB(3);"1. URBAN."
540 PRINT TAB(3);"2. SUB-URBAN."
550 PRINT TAB(3);"3. RURAL."
560 PRINT
570 INPUT "(1,2 OR 3)";P
610 IF P<3 AND P>1 GOTO 620

615 GOSUB 3000
616 GOTO 570
620 PRINT CHR$(147);"Q4. :SUN/SHADE/ORIENTATION."
630 PRINT TAB(3);"1. FULL SUN."
640 PRINT TAB(3);"2. SLIGHT SHADE."
650 PRINT TAB(3);"3. HEAVY SHADE."
660 PRINT TAB(3);"4. N. OR E. FACING WALL."
680 PRINT
690 INPUT "(1,2,3 OR 4)";H
710 IF H<4 AND H>1 GOTO 720

715 GOSUB 3000
716 GOTO 690
720 PRINT CHR$(147);"Q5. :SOIL TYPE."
730 PRINT TAB(3);"1. CHALK OR ALKALI."
740 PRINT TAB(3);"2. SAND."
750 PRINT TAB(3);"3. ORDINARY LOAM."
760 PRINT TAB(3);"4. CLAY OR ACID."
770 PRINT
780 INPUT "(1,2,3 OR 4)";S
790 IF S<4 AND S>1 GOTO 800

795 GOSUB 3000
796 GOTO 780
800 B=0
810 PRINT CHR$(147);"CHECKING THE DATA LIST"
820 FOR I= 1 TO T
```

Operation

470 Answer question.
480 Check answer. Skips next two lines if correct.
490 To error sub-routine if incorrect.
500 Returns to ask question again.
520—550 Third question.

570 Answer question.
610 Check answer. Skips next two lines if correct.
615 To error sub-routine if incorrect.
616 Returns to ask question again.
620—660 Fourth question.

690 Answer question.
710 Check answer. Skips next two lines if correct.
715 To error sub-routine if incorrect.
716 Returns to ask question again.
720—760 Fifth question.

780 Answer question.
790 Check answer. Skips next two lines if correct.
795 To error sub-routine if incorrect.
796 Returns to ask question again.
800 Sets indicator to zero.

820 Sets up loop to check list.

Program 5 Contd.

Operation

Program

```
850  IF A(I,E)=1 GOTO 1000
920  GOTO 1530
1000 IF A(I,5+M)=1 GOTO 1100
1010 GOTO 1530
1100 IF A(I,8+P)=1 GOTO 1200
1110 GOTO 1530
1200 IF A(I,11+H)=1 GOTO 1300
1210 GOTO 1530
1300 IF A(I,15+S)=1 GOTO 1500
1310 GOTO 1530
1500 B=B+1
```

1500 Suitable type found, increments indicator by 1.
1502 To sub-routine when first suitable tree found.
1506 Prints name of tree or shrub.
1530 End of loop.
1800 If B=0 no suitable trees found.

```
1502 IF B=1 GOSUB 3100
1506 PRINT "",""," ,N$(I)
1530 NEXT I
1800 IF B<>0 GOTO 6000
2100 PRINT "NO SUITABLE TREES IN MY DATA BANK."
2200 PRINT "WOULD YOU LIKE TO TRY AGAIN"
2210 GOTO 6010
```

```
3000 PRINT "ERROR IN ENTRY-TRY AGAIN"
3001 RETURN
```
3000—3001 Error sub-routine.

```
3100 PRINT CHR$(147);"POSSIBLE TREES:-"
3110 PRINT
3120 RETURN
```
3100—3120 Sub-routine to print heading.

```
4000 REM DATA BANK
4005 FOR K=1 TO T
4010 READ N$(K)
4020 NEXT K
```
4000—4300 Data sub-routine.
4005 Sets up loop to read data (names of trees).

```
4030 DATA "ASH","BEECH","BIRCH","CEDAR","ELM","FIR"
4031 DATA "HAWTHORN","HOLLY","HORSE-CHESTNUT","JUNIPER"
4040 DATA "LABURNAM","LIME","MAPLE","OAK","PINE","PLANE"
4050 DATA "POPLAR","PRUNUS","RHUS","ROBINIA","ROWAN"
4060 DATA "SPRUCE","SYCAMORE","WILLOW","YUCCA"
4100 FOR K=1 TO T
4105 FOR M=1 TO 19
4110 READ A(K,M)
4115 NEXT M,K
4120 DATA 0,1,1,1,1,1,0,1,1,1,0,0,0,1,0,1,1
4125 DATA 0,1,1,1,1,1,1,0,1,1,1,1,1,1,0,1,0
```

4100 Sets up second loop to read data (suitability table).

Program 5 Contd.

Program	Operation

```
4130 DATA 1,1,1,1,1,1,1,1,1,0,1,1,1,1,1,1,1,1,1,0
4135 DATA 0,1,1,1,1,1,0,1,1,1,1,1,1,1,1,1,1,1,1,1
4140 DATA 0,0,1,1,1,0,1,1,1,1,1,1,1,1,1,1,1,1,1,0
4145 DATA 0,0,1,1,1,1,0,1,1,1,1,1,1,1,1,1,1,1,1,1
4150 DATA 1,1,1,1,1,1,1,1,0,0,1,1,1,1,1,1,1,1,1,0
4155 DATA 0,1,1,1,1,1,1,1,1,1,1,1,1,1,1,1,1,1,1,1
4160 DATA 0,1,1,1,1,1,0,1,1,1,1,1,1,1,1,1,1,1,1,0
4165 DATA 1,1,1,1,1,1,1,1,1,1,1,1,1,1,1,1,1,1,1,1
4170 DATA 0,0,1,1,1,1,0,1,1,1,1,1,1,1,1,1,1,1,1,0
4175 DATA 0,1,1,1,1,1,0,1,1,1,1,1,1,1,1,1,1,1,1,0
4180 DATA 0,1,1,1,1,1,1,1,1,1,1,1,1,1,1,1,1,1,1,0
4185 DATA 1,1,1,1,1,1,1,1,1,1,1,1,1,1,1,1,1,1,1,0
4190 DATA 1,1,1,1,1,1,1,1,1,1,1,1,1,1,1,1,1,1,1,0
4195 DATA 0,1,1,1,1,1,1,1,1,1,1,1,1,1,1,1,1,1,1,0
4200 DATA 0,1,1,1,1,1,0,1,1,1,1,1,1,1,1,1,1,1,1,0
4205 DATA 0,1,1,1,1,1,1,1,1,1,1,0,1,1,1,1,1,1,1,0
4210 DATA 0,0,1,1,1,1,1,1,1,0,1,1,1,1,1,1,1,1,1,0
4215 DATA 0,0,1,1,1,1,1,1,1,1,1,0,1,1,1,1,1,1,1,0
4220 DATA 1,1,1,1,1,1,1,1,1,1,1,1,1,0,1,1,1,1,1,0
4225 DATA 1,1,1,1,1,1,1,1,1,0,1,1,1,1,1,0,1,1,1,1
4230 DATA 1,1,1,1,1,1,1,1,1,1,1,1,0,0,1,1,1,1,1,1
4235 DATA 1,1,1,1,1,1,1,1,1,1,0,1,1,1,1,1,0,1,1,1
4240 DATA 0,0,1,1,1,1,1,0,1,1,1,1,1,1,1,0,1,1,1,0
4300 RETURN
6000 PRINT "WOULD YOU LIKE TO TRY WITH A DIFFERENT"
6005 PRINT "SITE"
6010 INPUT "(Y OR N)";D$
6020 IF D$="Y" GOTO 310

6030 IF D$="N" GOTO 9000
6040 GOSUB 3000
6050 GOTO 6010
9000 END
```

6010 Opportunity to return to start.
6020 Transfers to beginning if answer is 'Y'.
6030 Transfers to end if answer is 'N'.
6040 To error sub-routine if incorrect.
6050 Returns to ask question again.

CHAPTER 9
File operations

If a data handling program is to be of any use to a design team it must contain a large amount of data. The problem of supplying a computer with data and hence producing a suitable data base has been discussed in the earlier chapters. In order to store a data base of any size it is necessary to introduce peripheral devices — these were discussed in section I and the application of the relevant devices will be dealt with in more detail in this chapter.

This however introduces a problem, as the method of accessing and using these peripherals varies from one machine to another. For this reason we have used the commands applicable to one particular machine, the Commodore PET. These commands may not be suitable for your machine and the equivalent commands can be found in your manufacturer's handbook.

In section II we have not previously referred to these devices and have assumed that all we have available is:

1. A keyboard, which is our INPUT device and enables us to key in a program or to key in replies to questions which have been programmed to appear on the screen.
2. The screen, which is our OUTPUT device which receives information from the central processor and displays it for our use.

In the central processor there are two types of memory and these were referred to in chapter 2. The portion of memory which can be accessed by the programmer is called a random-access memory (RAM) and is of the kind which will only retain information as long as a power supply is available. This means that any program or data which has been fed into the machine will be lost as soon as the main power supply is turned off. The machine also contains read-only memory (ROM) which contains the BASIC language compiler or interpreter etc. This memory is known as 'firmware' and is unaffected when the power is turned off. As its name suggests it cannot be used by the programmer as a storage medium.

To overcome the problem of losing programs or data when the power is turned off, some additional device is needed which will store this information. The simplest of these devices is the cassette tape which permits the storage of information in its magnetic coating in the same way that sound is stored on a similar audio tape.

The other method of storage is the magnetic disk which uses the same principle and which may be a solid disk fixed in a cabinet, or more likely with a microcomputer, a flexible diskette of plastic with a magnetic surface coating. The coating is similar to that used on a cassette tape.

Both the tape and the diskette can be either INPUT or OUTPUT devices, depending on how they are used. If we have typed in a program and wish to SAVE it for reuse later, the machine is instructed to SAVE the program on the tape or the diskette. In this case the cassette or disk is being used as an OUTPUT device because we are outputting information to it. When we wish to reuse the program, the machine is instructed to find the program and LOAD it so that in this case the cassette or disk is being used as an INPUT device.

Using a cassette tape is usually easier to understand and therefore is a good way to start. It is essential that the manufacturer's instructions are followed exactly, otherwise errors will occur. A cassette tape is perfectly effective as an INPUT or OUTPUT device — the disadvantage being its lack of speed and the problem that sometimes arises when the record/playback heads on one tape deck do not align with those of another machine even of the same manufacture. This can prevent programs written on one machine being loaded into another. The tape, as in any cassette recorder, moves past the record and playback heads at a fixed speed when recording or playing back and it takes several minutes for a tape to wind from one end to the other. A number of programs may be stored on one tape, so that if the program you wish to load is near the end and the tape is a long one, it can be several minutes before the machine finds the program you require. The length of time taken to load the program depends, of course, on the length of the tape taken to store it which in turn depends on the length of the program. A further disadvantage of the cassette tape is that it is normally quite easy to overwrite existing information on the tape, so destroying it. This is convenient when you wish to do this, but very frustrating when it occurs accidentally and ruins perfectly good information.

The disk drive operates by rotating the disk and moving a read/ record head radially. The disk is not read from start to finish to search for a particular item, but is divided into tracks which in turn are divided into blocks or sectors, each of which is referenced. By giving the correct instruction, a particular track and a particular block or sector on it can be read. As the disk rotates, its surface is scanned by the read/record head at a much greater speed than is the case with the cassette tape. This coupled with the ability of the disk drive to locate the portion of the disk where the required information is stored leads to a very short access time.

So that the disk drive can locate where a particular piece of information has been stored the disks are provided with a directory or
catalogue in which the names of the programs etc. are listed. When a
program is saved on disk, the computer makes an entry in the disk
catalogue which records the position of the program on the disk
together with its name. When we require a program to be loaded, the
system first reads the catalogue to see whether the program exists.
If it does, the head is moved to the starting sector of the program
and the program is loaded. Using a catalogue in this way removes
the danger of accidentally overwriting wanted information. A special
procedure is required when you wish to update a program before
it can overwrite the previous version. It is possible to dispense with
the catalogue system and make your own program keep track of
where the information is stored, but this is not recommended for
a beginner.

When information is stored on disk or tape it is said to be stored
in a file. If the information stored is a program it is called a PRO-
GRAM FILE and if the information is data, a DATA FILE.

The computer when first switched on has a series of default
values built in, so that it will expect information to be supplied from
the keyboard and will output the results to the screen. The commands required to change these default values will depend on the
type of machine, so that the manufacturer's instructions must be
followed exactly if you are to be successful.

The various pieces of equipment which make up the computer
installation, for example the screen, keyboard, tape drives, printers
etc. are all assigned their own device number so that each unit has
its own unique address. If we wish to output to a tape or disk drive,
an instruction must be given to the processor which informs it of
the device address which is to be used. So that several input/output
addresses may be stored at one time, a section of memory is allocated to this purpose. This section of memory is divided into several
channels, each of which has its own particular number. A channel
is a small section of memory which can be addressed by a program
statement using the channel number and into which we can place the
address of some peripheral device. In the case of the Commodore
PET the command OPEN is used to access a channel, so that

30 OPEN 2,1

will set the device address in channel number 2 to 1 which is the
device address of a tape drive. A secondary address may also be given
which denotes the type of operation that channel is to be used for.

30 OPEN 2,1,1,"FILE A"

means that the channel is to be used for writing data to the tape and the file name to be accessed is "FILE A".

This method of opening a channel applies to the Commodore PET and does not apply to other makes of microcomputer. Your own instruction book must be studied to find the equivalent commands.

Once a channel has been assigned in this way it forms the link between the processor and the external device and remains active until CLOSEd, the program is cleared or the machine is switched off.

Program files

It is usually possible to save a program which is stored in the computer's memory, by transferring it to a tape simply by typing in the command SAVE (RETURN). To recover the program from the tape LOAD (RETURN) will cause the program to be loaded into the memory of the machine. It is, of course, necessary to rewind the tape first. Programs which are saved on tape can be given a name on some computers, at the time the program is SAVEd. The maximum length of this name will depend on the machine. The advantage of naming a program which is saved on a tape is that the name can be used when you wish to reload the program. The computer will then search through the tape until the right program name is found and then load it. Any other programs that may be encountered are ignored. Issuing a LOAD command with no program name will cause the first program encountered to be loaded.

When a file is created by typing in SAVE (RETURN), the computer first writes a 'header record' which indicates that a program file is to follow. This 'header record' will include the file name if one has been given. The program itself follows and when this is complete the computer writes a 'trailer record' to denote that the file is complete.

Unless the protect tabs are removed from the cassette after a program is saved, it is possible to reuse the tape to save another program, so destroying the original one. A problem sometimes occurs when a number of programs are stored on a single tape and it is desired to modify one of them and then to resave it over its original version. If the length of the program is increased by this modification it is always possible that the 'header record' of the next file will be overwritten which will make that file inaccessible.

When a program is saved on disk, it *must* be given a name and this name must be used to retrieve the file. The system enters the program name in the directory or catalogue on the disk, provided that the name has not already been included. If the name already appears an error is signalled. Otherwise the system proceeds to store the program in much the same way that it was stored on tape, but this time its location on the disk is recorded against the program name in the catalogue.

If the program is later revised and the new version is to be saved in place of the old, some of the microcomputers require a special procedure to be adopted. The system will ensure that no other catalogued file is affected by this operation.

The storage capacity of the disk or tape will be much greater than the memory capacity of the computer so that program length is limited by memory size and not by the tape or disk.

Data files

A file which contains the values of variables is known as a data file. A file such as this cannot be accessed by simply using the command LOAD because the machine must also be given the variables into which the data is to be placed. It follows therefore, that a program must be written to provide this information and similarly a program is needed to put the data in the file in the first place.

As far as the computer is concerned, data files can be in three forms:

1.   Sequential storage files:

   (a)   Serial;
   (b)   Sequential.

2.   Relative files.
3.   Random-access files.

Using the cassette tape as the means of storing the data, only sequential storage files are possible and this type of file will be dealt with first. The other two are suitable when disk storage of the data is to be used.

Sequential storage data files

When discussing file arrangements in section I (chapter 5), the first two arrangements mentioned were 'serial files' and 'sequential files'. The only difference between these two types is the order in which the data is stored within the file — a serial file contains data in a random order and a sequential file contains sorted data. As far as the machine is concerned, both arrangements use sequential storage — that is, the records which are stored in the file follow one another directly and these records may be of varying length.

As described earlier in this chapter, a channel of communication must first be set up between the processor and the disk drive and in the case of some microcomputers this will also apply to the cassette tape. In the case of the Commodore PET, any statements which require to access the tape or disk drives include the address # N where N is the chosen slot number. # indicates that the number which follows is a channel number.

In the case of the Commodore PET the statements PRINT, IN-PUT and GET can all be used for writing to, or obtaining data from, tape or disk. In the case of the Apple these statements apply to the disk drive only.

20 PRINT A$

prints the value of A$ on the screen. The following lines:

10 OPEN 1,1,1,"FILE A"
20 PRINT # 1,A$
30 CLOSE 1

apply to the Commodore PET and write the value of A$ to the tape using channel 1. Line 10 opens channel 1 to the tape drive, the third parameter indicates that it is to be used for writing to the tape and the fourth parameter gives the name of the file which is to receive the data. Line 20 causes the value of A$ to be written to the tape and line 30 closes the channel. This little program creates a 'logical record' on the tape which has one 'field' — that is, the value of one variable is stored in the record. The meanings of 'logical record' and 'field' have been discussed in chapter 5.

Dealing now with the case where we require to retrieve the value of A$

20 INPUT A$

allows the value to be typed in using the keyboard.

10 OPEN 2,1,0,"FILE A"
20 INPUT # 2,A$
30 CLOSE 2

opens channel 2 to the tape drive. The third parameter is set to zero which indicates that the channel is to be used for reading from the tape, the fourth parameter giving the name of the file which is to be accessed as "FILE A". Line 20 directs that the value of A$ is to be read from the data file "FILE A" on the tape using channel 2. Line 30 closes the file.

The GET statement can also be used to read data, but in this case the data is read one character at a time and not as a complete variable.

In the case of the Apple, data files on tape are not supported to any great extent and the method used to access disk data files varies with the BASIC being used. In general, once a data channel has been opened to a disk by specifying the necessary parameters, the statements PRINT, INPUT and GET will automatically access the disk and not the screen.

When information has to be transferred from the processor to the tape or disk, a buffer is used. This is a section of memory which may be in the processor or in the peripheral device, into which the data is placed as it is generated by the program. When the buffer is full the contents are transferred to the tape or disk. In the case of the Commodore PET and the Apple, this information could remain in the buffer if it is not full when the transfer operation is complete. The CLOSE statement prevents this from occurring, as one of its functions is to flush the contents of the buffer to the output device.

The buffer is also used for transfers from an output device to the processor. A batch of data is read and placed in the buffer from where it is directed to the correct locations in memory.

The previous examples have all placed the value of a single variable in a file, thus creating a logical record containing a single field. It is more usual to require a file to hold many data values so that a single record may contain many fields.

20 PRINT # 1;A$;B$;C$;D$

will create a logical record which one might expect to contain four fields. The method used by the machine to store this data on tape or disk varies. Some machines will automatically place a marker at the end of each variable when it is transferred to the buffer, in which case the record will contain the expected four fields. In the case of most of the microcomputers, however, when the variables are transferred to the buffer, B$ follows immediately after A$, C$ after B$ etc., without a division of any kind so that there is no record of the original length of these variables and the record contains only a single field. In order that the variables can be recovered correctly a divider must be used by the programmer.

20 PRINT #1;A$;",";B$;",";C$;",";D$

will insert a comma between each field which will provide the necessary separation. If the variables to be saved are numeric, the Commodore PET and the Apple require the spacers to be carriage returns. Carriage returns are also permissible as spacers for alphanumeric variables. The way in which a carriage return is inserted to form a field divider is by using CHR$(13) which is the code meaning a carriage return. The line would now read:

20 PRINT # 1,A$;CHR$(13);B$;CHR$(13);C$;CHR$(13);D$

When it is desired to obtain data from a file, the line:

120 INPUT # 1,A$,B$,C$,D$

is the comparable statement. If the markers have not been added in the PRINT # line, the values of A$, B$, C$ and D$ will all be read into A$ as though they had been concatenated. The variables B$, C$ and D$ will have no value.

If at some time we wish to add an additional variable to the file, it is not possible to do this directly. That is we cannot direct the system to the end of the record and add on the new value. The method which must be used is to first of all read all the existing data as in line 120 above, rewind the tape if tape is the storage medium being used, reopen the file and resave the data with the new value included. Suppose that E$ is the variable which is to be added to this file; then:

150 PRINT # 1,A$;",";B$;",";C$;",";D$;",";E$

will resave the original data with the added value at the end so that the logical record will now contain five fields.

Note that if an array is being saved, a FOR-TO-NEXT loop is used (see program 4(d), lines 360—400). The line:

370 PRINT # 1,K$(I)

causes the machine to record a carriage return if the machine is a Commodore PET and cassette tape is the storage medium. In this case it is not necessary to force a divider. This does not apply to all machines and does not apply even to the PET if the disk is being used to store the data. In this case the line would be written:

370 PRINT # 1,K$(I);CHR$(13)

The facility is provided in some systems, which place a marker at the end of each record, to skip over complete logical records within a file using SKIP and BACKSPACE. When this facility is available a data file can contain a number of logical records, each containing any number of fields. If it is not required to access the whole of the file, the desired record can be found, loaded into memory, updated and returned to its original position in the file.

In the case of the Commodore PET and the Apple, however, this is not possible. The reason is that no special marker is placed at the end of each record; also, as the lengths of the variables which form the individual fields can vary, the length of each record can be different. It is therefore impossible to identify the end of a record either by marker or by each record being of some fixed length.

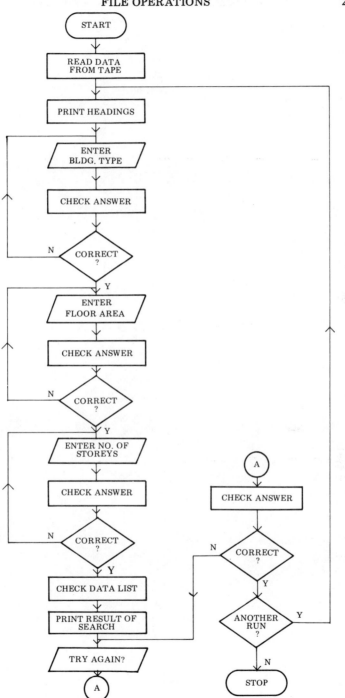

Fig. 9.1 Flow chart for program 4(b): search records 2.

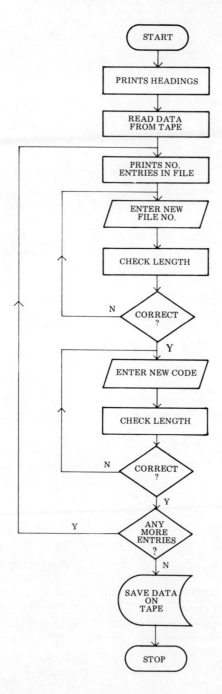

Fig. 9.2 Flow chart for program 4(d): data file access.

PROGRAM 4(b)

Chapter 7 introduced a program designed to search records (program 4(a)) in which the data was held in a series of DATA statements. Program 4(b), 'Search records 2' (see end of this chapter), is a modification of this earlier program in such a way that the data is now stored on tape and not written into the program (fig. 9.1). The program is written to suit the Commodore PET and the method of accessing the tape will have to be varied to suit other systems.

The new program is only modified slightly in that:

(1)   The lines 25—60 which read the data are replaced by the lines 25—65 which open the file and read the data from tape.

(2)   Lines 700—730 which contained the data are deleted.

There are however two additional programs, one of which — program 4(c), 'Create data file' — is a temporary program used to establish the data file on the tape (see end of this chapter). Once this has been done this program can be modified and extended into program 4(d) (see end of this chapter) which is designed to provide access to the data file in order that new entries can be made (fig. 9.2).

SOME DEVELOPMENTS

You may like to try to combine programs 4(b) and 4(d) into one single program. It will now be necessary to introduce a choice at the beginning which would determine whether the program was to search the records as did the original program 4(b) or to access the file to insert new data.

Program 4(d) requires the input of the three-digit code in line 270 in accordance with the system given at the beginning of chapter 7. Program 4(b) includes a routine which prints a series of questions on the screen, the answers to which are used to produce the code. If we combine the two programs as suggested above, we can use this routine to set up the code either to insert new data into the file or to search the data file for that particular code. If this is done, the operator need have no knowledge of the code which is being used.

To direct the route through the program, a variable, say D1, can be set by the answer to the question introduced at the beginning of the combined program. Using program 4(b) as the basis, the heading in lines 70 and 80 can be extended to describe the second function of the program, followed by the question asking which particular run is required. The answer to this question can be a numeric one used to set the variable D1 to 1 or 2. Suppose that if the run is to search the records as the original program 4(b) then D1=1 and if access to the files is required then D1=2.

If these values have been set in line 90,

   92 IF D1=1 THEN 100

will bypass the lines 93—99 if the original file search is required. This provides a space which can be used to accommodate the text of the lines 200—250 from program 4(d) which permit the entry of a file number. The rest of the program 4(d) can then be modified in a similar manner.

Relative data files

Most versions of the BASIC language used by the present-day micro-computers support the use of this type of file, provided that the storage medium is the disk. In this type of file, all the records are of the same length so that it is possible to calculate the sector on which a particular record and field within that record will be stored and therefore to access it, update it and return it to that sector without reading in any other part of the file.

Random-access files

Mention was made earlier in this chapter of the method used for transferring information between the processor memory and the tape or disk using a buffer as an intermediate store, where the data is held until the buffer is full. The normal size of this buffer is 256 bytes which is the same size as one sector on the disk. Therefore data which is collected in the buffer is exactly the correct amount to fill one sector, or conversely when reading a disk, the data from one sector will exactly fill the buffer. If the disk catalogue is being used to control the data file, not all the 256 bytes will be available to the programmer as two or three will be required as control bytes by the system.

Your instruction manual will give information concerning the number of tracks on your particular disk system, together with the number of sectors contained in each track. The Commodore PET (except those using BASIC 4.0) requires both track number and sector number to be specified in order to access a particular sector. As the number of sectors per track is not constant but increases as the tracks get longer towards the perimeter of the disk, careful programming is required. The 4000/8000 series Commodore PET and some other systems number the sectors from zero upwards and only require the sector number to be specified. The system is then able to locate the required track and the sector on it.

This facility of being able to address an individual sector makes it possible for the programmer to fill the buffer with data in any form he likes and then transfer it to a particular sector on the disk. Of course, it is the programmer's responsibility to write a program which will access the sector and read the data back into the buffer when it is needed and also decipher the data which the buffer contains.

Using this method, it is possible to use a complete disk as a data file which is independent of the disk catalogue or directory. The program must calculate the sector required for any read or write operation.

Files of this kind are referred to as random-access files and a number of special statements are provided for accessing the disk in this way. *It is not recommended that this method be used until some experience has been gained in the use of sequential, catalogued files and programming in general.*

Reference is made at the end of the previous chapter to the situation where the amount of data stored in a file was too great to be stored in the computer memory all at the same time. The example referred to selected information provided by a technical index or similar system and the problem was concerned with a method of placing such information in some specific order. This was achieved by producing a locator array which held the required order; the actual data was not physically moved in the file.

If we consider the same problem again, but this time dealing with the file, a random-access type of file would be a suitable solution. One sector on a disk can hold 256 bytes and this may well be enough space to contain all the desired information about one manufacturer. (The Commodore PET only permits 254 of these bytes to be used by the programmer, but this is not the case with most other machines.) If all the information concerning one manufacturer can be stored on a single sector, each manufacturer can be allocated a particular sector on the disk. If one sector is not sufficient more may be used but this will make the calculation of the position of the sectors on the disk a little more complex. The size of the file will therefore depend on the number of sectors on the disk and how many are to be used for each record. This will vary quite considerably depending on the type of machine and disk drive being used. Remember that some sectors will have to be allocated to the storage of the locator array or arrays and any other control information which may be needed. If a double disk drive is available there is no reason why the file should not extend to both disks.

Suppose that one sector is to be used for each record and the whole of the 256 bytes on that sector are available to the programmer. 25 bytes are allocated to the name of the manufacturer, 100 bytes to the address and the remaining 131 to the product information. The program must pack out these variables to their correct length if the BASIC used does not permit the DIMensioning of a string to include its length.

A program can be written to permit the entry of new data in the usual way using INPUT statements. When the data concerning one manufacturer is complete, this data is allocated to a buffer. Because the variables which are to be transferred to the buffer are all of known lengths we do not need to place spacers between them and simple concatenation can be used. When the data is read by another program the contents of the buffer can be transferred into the correct variables using LEFT$(, STR(, SUB$(etc.

When the buffer has been filled, the data can be transferred to a vacant sector on the disk. This means that the program must keep a record of the sectors which have been used. This does not present any particular problem if the sectors are used sequentially. A variable is used to store the number of the next available sector and this is incremented by one each time a new record is entered. This variable must also be stored in the file along with the locator array or arrays. The reference number of the entry is based on the sector number.

If provision is made for deleting entries and making the sectors available for new data then this may be done by using an array which contains the numbers of any such sectors. When deletion occurs the program enters the number of the sector concerned in this array. Subsequently when new data is entered, the program must search this array to check for vacant sectors and if any are found, to reuse them in preference to a new sector at the end of the file. The program then removes this sector number from the array.

Another program must then be written which will sort the data which has been entered. This program allows for the selection of a key, for example the names, products etc. Using a FOR-TO-NEXT loop every sector containing data is loaded into the memory in turn, the desired key extracted and placed in a sort array. When the sort array is complete, the 'sort' routine, given in the previous chapter, can be used to produce the locator array. This sort program can be made to follow automatically whenever new data has been added so that an up-to-date locator array is always available.

A third program can now be written which will print out the sorted data in whatever form is required. The data may be printed either on the screen or using a printer. This program will use the locator array produced by the sort program to access the sectors on the disk in the correct order. Each sector containing data is loaded into the memory in turn and then printed to whatever format is required. It is useful to arrange for the printing of an alphabetical list of names together with the reference numbers (based on the sector numbers), which can then be used to access a particular record for revision.

The fourth program in this series is one designed to read a particular record from the tape, allow it to be revised and then return it to the same sector. The record is found by the system using the reference number typed in as an answer to a question.

This example on the use of the random-access method has been limited to a description only, because the methods of putting it into use on different machines will require different programming methods and command statements. It is therefore essential to understand thoroughly the method of accessing the disk in this way for your particular machine.

Program 4(b) Search records 2.

Program

```
10 PRINT CHR$(147)
20 DIM K$(100),K(100)
25 PRINT "MOUNT TAPE-TOUCH RETURN WHEN READY
30 GET X$:IF X$="" THEN 30
35 OPEN 1,1,0,"DATFIL"
40 FOR I=1 TO 100
45 INPUT #1,K$(I)
50 NEXT I
55 INPUT #1,V
60 CLOSE 1
65 PRINT

70 PRINT "THIS PROGRAM WILL SEARCH RECORDS"
80 PRINT "TO FIND SIMILAR TYPES OF JOB"
90 PRINT
100 PRINT "TYPE OF BUILDING"
110 PRINT TAB(3);"1) SCHOOL"
120 PRINT TAB(3);"2) FACTORY"
130 PRINT TAB(3);"3) DOMESTIC"
135 N=0
140 INPUT "(1,2 OR 3)";N
150 D=3
160 GOSUB 1000
165 IF D=1 THEN 135

170 NO=N
180 IF NO=3 THEN GOSUB 1500
190 PRINT CHR$(147);"FLOOR AREA"
195 PRINT
200 PRINT TAB(3);"1)    0-99   SQ.M"
210 PRINT TAB(3);"2)  100-199  SQ.M"
220 PRINT TAB(3);"3)  200-299  SQ.M"
230 PRINT TAB(3);"4)  300-399  SQ.M"
240 PRINT TAB(3);"5)  400-499  SQ.M"
250 PRINT TAB(3);"6)  500+     SQ.M"
255 N=0
260 INPUT "(1,2,3,4,5 OR 6)";N
270 D=6
```

Operation

10 Clears screen.
20 Dimension arrays.
25 Prints message.
30 Waits until ready.
35 Opens the data file on tape.
40—55 Reads the data from tape.

60 Closes the data file.
65 Leaves blank line on screen.
Used throughout program.
70 & 80 Print heading.

100—130 Sets up first question.

135 Sets variable used in question to zero.
140 Answer question.
150 Sets error variable to maximum value.
160 To sub-routine to check answer.
165 Return to ask question again if incorrect.

180 To sub-routine if domestic.
190—250 Second question.

255 Sets variable used in question to zero.
260 Answer question
270 Sets error variable to maximum value.

Program 4(b) Contd.

Operation	Program
280 To sub-routine to check answer. 285 Return to ask question again if incorrect.	```280 GOSUB 1000``` ```285 IF D=1 THEN 255```

300 Third question.	```290 N1=N```
	```300 PRINT CHR$(147);"NUMBER OF STORIES"```
	```310 PRINT```

315 Sets variable used in answer to zero.
320 Answer question.
330 Sets error variable to maximum value.
340 To sub-routine to check answer.
350 Return to ask question again if incorrect.
	```315 N=0```
	```320 INPUT "(1,2,3,4,5 OR 6)";N```
	```330 D=6```
	```340 GOSUB 1000```
	```350 IF D=1 THEN 315```

360—380 Set A$ to suit answers to above questions.	```355 N2=N```
	```360 A$=MID$(STR$(N),2,1)```
	```370 A$=A$+MID$(STR$(N1),2,1)```
	```380 A$=A$+MID$(STR$(N2),2,1)```

390—410 Search loop.	```390 FOR I=1 TO V```
	```400 IF A$=RIGHT$(K$(I),3) THEN K(I)=1```
	```410 NEXT I```

420 Prints heading.	```420 PRINT STR$(147);"RESULT OF SEARCH"``` ```430 PRINT```

435 Sets D to zero — the value which means 'no similar jobs'.
440—465 Loop to print results.
450 Prints file number if K(I)=1.
	```435 D=0```
	```440 FOR I=1 TO V```
	```450 IF K(I)=1 THEN D=1:PRINT "FILE NUMBER",LEFT$(K$(I),8)```

460 Resets K(I) to zero.	```460 K(I)=0``` ```465 NEXT I```

470 If D is still zero there are no similar jobs on file.
490 Prints question.
495 Sets variable in answer to zero.
500 Answer question.
	```470 IF D=0 THEN PRINT "NO SIMILAR JOBS FOUND"```
	```490 PRINT "DO YOU WISH TO SEARCH FOR MORE"```
	```495 Q=0```
	```500 INPUT "1) YES, 2) NO";Q```
	```510 ON Q GOTO 540,550```
	```515 N=10```
	```520 GOSUB 1000```

520 To error sub-routine if answer incorrect. 500 Return to ask question again. 540 Clear screen.	```530 GOTO 500``` ```540 PRINT CHR$(147)```

Program 4(b) Contd.

Program

```
 550 GOTO 100
 560 END
1000 IF N(1 OR N)D OR N()INT(N) THEN PRINT "ERROR
     IN ENTRY-TRY AGAIN":D=1:GOTO 1020
1010 D=0
1020 RETURN
1500 PRINT STR$(147);"TYPE OF DOMESTIC BUILDING"
1505 PRINT
1510 PRINT TAB(3);"1) DETACHED HOUSE"
1520 PRINT TAB(3);"2) SEMI-DETACHED HOUSE"
1530 PRINT TAB(3);"3) TERRACED HOUSE"
1540 PRINT TAB(3);"4) FLATS"
1545 N=0
1550 INPUT "(1,2,3 OR 4)";N
1560 D=4
1570 GOSUB 1000
1580 IF D=1 THEN 1545

1590 N0=N0+N

1500 RETURN
```

Operation

550 Return to start of program.
560 Program run stops.
1000 Error sub-routine.

1500 Sub-routine to choose type of domestic building.

1545 Set variable used in question to zero.
1550 Answer question.
1560 Sets error variable to maximum value.
1570 To sub-routine to check answer.
1580 Return to ask question again if incorrect.
1590 Revises value of N to suit answer to last question.

Program 4(c) Create data file.

Program

```
10 PRINT CHR$(147)
20 DIM K$(100)
30 PRINT "MOUNT TAPE-TOUCH RETURN WHEN READY"
40 PRINT

50 GET X$:IF X$="" THEN 50
60 PRINT "OPENING FILE"
70 OPEN 1,1,2,"DATFIL"
80 FOR I=1 TO 100
90 PRINT#1,K$(I)
100 NEXT I
110 PRINT#1,V
120 CLOSE 1
130 PRINT
140 PRINT "FILE ESTABLISHED"
```

Operation

```
10 Clear screen.
20 Dimension variables.
30 Prints message.
40 Leaves blank line on screen. Used throughout program.

50 Waits until ready.
60 Prints heading.
70 Opens file.
80—120 Places blank data in file
```

Program 4(d) Data file access.

Program

```
10 PRINT CHR$(147)
20 DIM K$(100),K(100)
25 S$="        "
30 PRINT "MOUNT TAPE-TOUCH RETURN WHEN READY"
40 PRINT
50 GET X$:IF X$="" THEN 50
60 PRINT "OPENING FILE"
70 OPEN 1,1,0,"DATFIL"
80 FOR I=1 TO 100
90 INPUT#1,K$(I)
100 NEXT I
110 INPUT#1,V
115 CLOSE 1
130 PRINT "DATA LOADED-REWIND TAPE."
140 PRINT"TOUCH RETURN WHEN READY" 150 GET X$
160 PRINT CHR$(147)
170 PRINT "THERE ARE";V;" ENTRIES IN THE FILE"
180 PRINT "ENTRY NUMBER";V+1
200 F$=""
210 INPUT "FILE NUMBER";F$
230 L=LEN(F$)
240 IF L)8 PRINT "FILE NUMBER TOO LONG":GOTO200
250 K$(V+1)=F$+LEFT$(S$,8-L)
260 C$=""
270 INPUT "3 DIGIT CODE";C$
280 L=LEN(C$)
290 IF L)3 PRINT "CODE TOO LONG":GOTO 260
300 K$(V+1)=K$(V+1)+C$
310 V=V+1
320 INPUT "ANY MORE ENTRIES (Y OR N)";Q$
330 IF Q$="Y" THEN 160
340 IF Q$()"N" THEN PRINT "ERROR IN ENTRY":GOTO 320
350 OPEN 1,1,2,"DATFIL"
360 FOR I=1 TO 100
370 PRINT#1,K$(I)
```

Operation

10 Clear screen.
20 Dimension variables.
25 Set blank string.
30 Prints message.
40 Leaves blank line on screen. Used throughout program.
50 Waits until ready.
60 Prints heading.
70 Opens file.
80—120 Loads data from file.

150 Waits until ready.
160 Clears screen.
170 Prints heading.
180—320 Enter new data.

320 Allow return for more entries if required.

350 Open file to save data.
360—400 Save data.

Program 4(d) Contd.

Operation

Program

```
380 NEXT I
390 PRINT#1,V
400 CLOSE 1
410 END
```

CHAPTER 10

Bits, bytes and program efficiency

Bits and bytes

Chapter 2 introduced and explained the reason for the method which is used to communicate with a computer employing binary arithmetic. Counting in binary is as follows:

Decimal		Binary	
Decimal	0	Binary	0000
	1		0001
	2		0010
	3		0011
	4		0100
	5		0101
	6		0110
	7		0111
	8		1000
	9		1001
	10		1010
	11		1011
	12		1100
	13		1101
	14		1110
	15		1111

It was explained in chapter 2 that the 'switches' in a computer are known as 'bits' and are grouped together — normally in sets of eight to make one 'byte'. The above example shows a grouping of four bits which can represent any decimal number from 0 to 15. If we use eight bits and continue as above, the eight bits can represent any number from 0 to 255.

Each digit position in a binary number represents an increasing power of 2. If a zero is used in a position you do not take the value of that position, if a one is used you do take the value of the position.

The letters of the alphabet, the decimal digits and the punctuation marks are all allotted their own particular number using a code. The code normally used is called ASCII (American Standard Code for Information Interchange). A carriage return in this code has a decimal value of 13, a space has a value of 32, zero has a value of 48, the

numbers then follow as 49, 50 etc. representing 1, 2 etc. Letters of
the alphabet start at 65 with the letter A. In chapter 8, when dealing
with the sorting of alphanumeric data, reference was made to the
values of letters and digits etc. It is these values that the computer
compares during a sort operation. The various BASIC words are
also represented in the same way, so that generally speaking a BASIC
word requires only one byte of memory. The codes used for the
BASIC words and any characters outside the ASCII code are not
standard and therefore vary from one make of computer to another.

The following diagram shows the value of each bit and how a
byte (or word in the case of any computer which uses an eight-bit
word) is built up. The diagram shows how the letter A would appear
in memory.

Power of 2	: 7 : 6 : 5 : 4 : 3 : 2 : 1 : 0 :
Value of position	:128: 64: 32: 16: 8 : 4 : 2 : 1 :
Binary digit	: 0 : 1 : 0 : 0 : 0 : 0 : 0 : 1 :
	: : : :
	: High order bits : Low order bits :

Most versions of BASIC permit easy conversion between codes and
the characters which they represent. The command CHR$( ) and
ASC( ) are used on the Commodore PET and the Apple.

PRINT CHR$(50)

will print a figure 2 on the screen. This method has been used several
times in the example programs in the earlier chapters of this book as
a method of clearing the screen. The code for this on the Commodore PET is 147 so that:

PRINT CHR$(147)

will clear the screen.

PRINT ASC("2")

will print the decimal value of the code for the figure 2, i.e. 50.

We have shown at the beginning of this chapter that four bits can
be arranged in a total of 16 different ways, representing the decimal
numbers 0—15. If we use a counting system that uses sixteen as a
base instead of the normal ten of the decimal system we can use this
as a shorthand way of expressing a binary value. The hexadecimal
counting system is such a system. In the decimal system we use
the digits 0—9. The hexadecimal system, however, requires the use
of fifteen digits and zero. The letters A to F are used as the extra
symbols and the digits used are 0, 1, 2, 3, 4, 5, 6, 7, 8, 9, A, B, C, D,
E and F. Combination of these digits, as with combinations of 0—9

in the decimal system, are used to represent all numbers. The comparison between decimal, binary and hexadecimal is as follows:

Decimal	0	Binary	0000	Hex	0
	1		0001		1
	2		0010		2
	3		0011		3
	4		0100		4
	5		0101		5
	6		0110		6
	7		0111		7
	8		1000		8
	9		1001		9
	10		1010		A
	11		1011		B
	12		1100		C
	13		1101		D
	14		1110		E
	15		1111		F

From this table we can see that the numbers represented by four bits can be written as a single digit using hexadecimal notation. As each byte of the computer's memory consists of 8 bits, it is convenient to split these into two groups of four and to refer to the four 'right hand' bits as 'low order' and the four 'left hand' bits as 'high order'. Continuing to count in this way we eventually reach the situation where both the low and high order bits reach their maximum value:

Decimal	16	Binary	0001 0000	Hex	10
	17		0001 0001		11
	18		0001 0010		12
	.				
	.				
	.				
	255		1111 1111		FF

One eight-bit byte can therefore have 256 values (including zero). Some versions of BASIC use hex values in place of the decimal numbers used by some of the microcomputers in the statements similar to CHR$( ) and ASC( ). It is also possible to manipulate the bits within a byte. The degree to which this may be done depends on the version of BASIC being used. Microcomputers support this to a limited extent using the Boolean operators AND, OR and NOT. In some versions of BASIC it is possible to add or subtract from the value of a byte — suppose A$="A" then the byte representing A$ is set to 0100 0001 = 65 decimal = 41 hex. If 1 is added to this it becomes 0100 0010 = 66 decimal = 42 hex which is the value of "B". If this kind of manipulation is possible a single byte can be used as a counter, provided that the maximum value of 255 is suitable for the particular purpose. If two bytes are used as a counter, then the maximum value becomes 65 535. If, on the other hand, a numeric variable is used as a counter it will require 7 or 8 bytes depending on the machine.

Saving of a few bytes of memory in this way is not important if only one counter is required. One of the methods used for sorting data in chapter 8 produced a locator array which contained the sequence in which the data was to be taken so that it would be in alphabetical or any other chosen order. To store the required sequence, a numeric array was used. This array was dimensioned as L(500) so that it would contain 500 elements. If we assume that the microcomputer uses 7 bytes to store a numeric value, the 500 elements of the array will require 500 × 7 = 3500 bytes of memory for its storage. The numbers which have to be stored in this array are all integers and range from 1 to 500 so that an element length of two bytes in a string array would be suitable to contain these values. The number of bytes of memory required would be 500 × 2 = 1000, a net saving of 2500 bytes.

If the BASIC language which is used by your computer allows for full manipulation of the bits within a byte, a considerable saving in memory requirements can be achieved in program 5, 'Trees' (see chapter 8). Looking at the DATA statements in lines 4120--4240 it can be seen that the data simply consists of 1 if the tree is suitable and 0 if it is not. The array used is numeric and has the dimensions of 30 × 20 which is 600 elements. Each element will require at least 7 bytes giving a total of 4200 bytes. If, however, one bit can be used for each data value, each byte will be capable of holding eight values and only three bytes will be needed to contain the whole of the data in one of the DATA lines. In fact three bytes will contain 24 values as against the 19 which appear on each DATA line of the program. The number of bytes now required to cater for up to 30 species of tree as before will be 30 × 3 = 90 bytes — a saving of 4110 bytes. Some of these bytes which have been saved will of course be used in the extra programming needed to set the bits to their correct value as the questions are answered; but against this, the loop which checks the list will be much simpler because the Boolean operator AND can be used to compare the string which is set by the answers to the questions, with the elements of the array containing the data. Boolean operators compare the bits within two strings one at a time; the effect of the AND operator is to set the bits in the result to 1 if the corresponding bits in the two strings are both 'on' and to zero in all other cases. Thus if the result of the AND statement is the same as the string set by the answers, the tree or shrub is suitable.

Program efficiency

When first starting to write your own computer programs, simply getting them to work seems quite an achievement. As experience is gained, the programs tend to become more complex and it is then time to consider the efficiency of your programs.

By program efficiency we mean

  (i) The minimum amount of memory space is used.
 (ii) The time taken to run the program is as short as possible.
(iii) The program is laid out in such a way and properly signposted with REM statements in order that it can easily be understood.
(iv) Errors in entry during the running of the program are detected and reported.

Dealing with these factors in turn, one way of reducing the memory requirement has already been mentioned earlier in this chapter but it is not a method which is possible on all machines.

In general the programs and starter pack programs in this book have been written with one statement per line for the sake of clarity. The only exceptions to this are:

(a) In the case of the IF-THEN statement where the syntax required by the Commodore PET and the Apple is that if the specified condition is true, the statement or statements on the same line are executed. If the specified condition is false, the program continues on the next line.
(b) To save space where a number of PRINT statements occur together or when values have to be assigned to a number of variables at a particular point in the program.

If your machine permits the use of multi-statement lines this facility should be used as much as possible. There are three reasons for this:

1. Permits logical grouping of statements. A complete FOR-TO-NEXT loop can often be contained in a single line.
2. The colon which is used to separate the various statements on a line requires one byte of storage in memory. A line number requires between two and five bytes depending on the system.
3. Program execution is faster.

Multi-statement lines will therefore both reduce memory requirements and increase the speed of processing. If multi-statement lines are permitted there is no limit to the number of statements which may appear on the same line. The limiting factors are the maximum length of line permissible on your machine and the fact that the GOTO statement can only be used to go to the beginning of a line and not to a statement within a line.

FOR-TO-NEXT loops usually take an appreciable time to process, so that if any alternative is available it should be used in preference. An example was given in chapter 7 of the use of MAT READ and if these statements are available they should always be used. The statements include MAT PRINT, MAT INPUT and MAT=, all of which take the place of a FOR-TO-NEXT loop.

It has been shown in the example programs in the earlier chapters that where a particular routine or section of a program is used several times, this can be in the form of a sub-routine. Using sub-routines in this way reduces the amount of memory which is needed to store the program as the routine appears only once but does not affect the running time of the program to any degree. The statements which have to be processed are only increased by two — the GOSUB and RETURN statements.

The first thing to do once your program has been typed into the machine, as mentioned at the beginning of section II, is to debug the program. The first stage of this process is to see if the program will run at all, and then, if it does, whether the questions appear correctly on the screen and whether incorrect answers to these questions are automatically rejected. This matter of the rejection of incorrect replies is dealt with in more detail later in this chapter. When checking the arithmetical calculations which you have included in your program, some test data must be fed in and the answers obtained by the machine checked against your own calculations in which you have used the same data. This is a tedious process but essential. To assist in this process it is useful to include some additional temporary PRINT statements so that interim results to calculations can be printed on the screen and checked against your own test calculations.

When programs become more complex, problems often arise over the use of variables. For example, the test run of a program may fail to give the correct answer to a particular problem, although reading through the program does not immediately reveal an error. A possible cause of this is that the same variable has been used for two different purposes in such a way that its value in some calculation has been changed from that intended. Reusing the same variable for several purposes in a program saves memory space but care is needed when writing the program to avoid the occurrence of this problem. It is good practice to use single variable names as control variables, i.e. X, Y, Z etc. and double variable names for actual values, i.e. X1, Y1, Z1.

In program 4(b) (see chapter 9), the variable D was used several times. Each time a question which required a numerical answer was asked, the variable D was set to the maximum possible answer to that question. The sub-routine in line 1000 then resets the same variable D to zero or one depending whether the reply given in answer to the question is a possible one or not. In cases such as this, the reuse of the same variable is not likely to cause any problem as the use is confined to a small routine in the program.

To avoid the problems which may occur due to the reuse of a particular variable it is good practice in all but the simplest of programs to compile a list of the variables which have been used and what they have been used for. When a new variable is needed, a glance at the list will show immediately which variables are still available.

The third aspect of program efficiency was that the program is laid out and properly signposted in such a way that it can be easily understood. The starter packs have been supplied with a liberal number of REM statements so that this object can be achieved. The programs contained within the chapters have been dealt with in a slightly different way in that a full description of the program operation has been included as a separate item alongside the actual program.

When writing more complex programs, it is essential that full documentation is produced as the program is written. In the authors' experience it has been found that not only should the program be provided with adequate REM statements and a list of variables and their uses compiled as described above, but, in addition, a written description of the various methods of programming which you have used should be compiled. Notes for this should be written as you write the program and revised as necessary as any alterations are made to the program. It is particularly useful to describe in detail how your data files have been compiled, how they are accessed and precisely what they contain. This kind of information can prove invaluable at a later date if any alterations or additions have to be made to your program.

It is quite easy to make mistakes when replying to questions which appear on the screen, so that it is good practice to build in a check routine to see whether the reply is a possible answer to the question. An example of this occurs in programs 4(a) and 4(b) where there is an error sub-routine on lines 1000–1020. In any program which requires a large amount of data to be entered via the keyboard it is useful to be able to edit any values which have been entered incorrectly. This can be done at the end of a series of questions and before the screen is cleared ready for the next series. An additional question is included — such as 'DO YOU WISH TO EDIT ANY OF THE ABOVE ENTRIES (Y OR N)'. Answering 'N' to this question allows the program to proceed. Otherwise the question 'EDIT WHICH' appears. For ease of reference at this stage, the questions on the screen should be numbered. Using the cursor control features of your machine, the cursor can now be directed to the required answer and the value typed in again. So that we do not have to answer all the questions again, an additional variable is introduced which is used to indicate when edit mode is being used. Suppose we select the variable E for this purpose and when E=1 an answer is to be edited, otherwise E=0. After each question the value of E must be tested and if its value is 1 the program jumps to the 'EDIT' question and if its value is zero the program continues with the next question in the normal sequence. If this is not done one mistake in entry will make it necessary to start the program from the beginning and re-enter all the data.

Screen presentation

No attempt has been made in the programs to produce a good screen presentation other than an occasional 'clear screen'. The methods of formatting the screen vary from machine to machine as do the screen sizes and graphics capabilities, so this is again a case for studying the manufacturer's instruction manual. The screen arrangement during the input of data or the printing of operating instructions is a matter of personal taste, the main object being clarity. To obtain this clarity the data input should preferably be in some definite position on the screen, perhaps with the answers to a series of related questions being placed in boxes. If an error in entry is discovered by the check routine in the program, a message can be displayed for a short period of time and then the incorrect answer removed from the screen so that the question is always answered in the same position.

To obtain a time delay we can use the real time clock if one is available on your machine. The Commodore PET uses the numeric variable TI as the clock which increments by one each 1/60th of a second. The following lines will produce a delay of one second.

```
100 T=TI
110 IF TI<T+60 THEN 110
```

If your program requires a number of time delays this line would become a sub-routine. If we use another variable in place of the figure 60 in line 110, this variable can be set before the sub-routine is entered. This makes it possible, if desired, for the delay time to be different on each occasion that the sub-routine is used.

If your machine does not have a real time clock, a FOR-TO-NEXT loop can be used as the delay. As mentioned earlier, a loop such as this takes some time to process so that a loop with no internal statements can be used as a time delay. The line:

100 FOR I=1 TO 100: NEXT I

will produce a short delay. The actual time taken and hence the length of the delay will depend on your machine. It could well occur that should you purchase a new machine in the future, the operation time of the FOR-TO-NEXT loop could be much reduced so that if any alternative method of producing a delay is available, it should be used in preference.

When a series of related questions have been answered successfully, it is a good plan to clear the screen and make a fresh start at the top for the next set of questions.

If an INPUT statement is used to receive a numeric value, e.g.

10 INPUT N

and an alpha character is inadvertently entered, an error message will be printed on the screen. This will, of course, upset your screen presentation. We can get over this problem simply by changing the line to:

10 INPUT N$

which will accept any value which is typed in on the keyboard and no error message will be produced by the machine.

The presentation on the screen has now been protected but we must check that the actual value typed in is a suitable answer to the question. This can be done in your program and if an error is discovered your own error routine can be used to print the error message and then the cursor moved to the correct position to permit re-entry of the incorrect value.

If we use programs 4(a) or 4(b) as an example, the first question (in lines 100—130 contains the type of building. The answer must lie between 1 and 3 so we set the variable D to the highest possible value for that answer (3) and branch to the sub-routine in line 1000 in order to test the validity of the input.

135 N$=" "
140 INPUT "(1,2 OR 3)";N$

```
145  N=VAL(N$)
150  D=3
160  GOSUB 1000
165  IF D=1 THEN 135
   .
   .
   .
1000 IF N <1 OR N>D OR N < > INT(N) THEN PRINT "ERROR
     IN ENTRY—TRY AGAIN":D=1:GOTO 1020
1010 D=0
1020 RETURN
```

In this version of the INPUT routine we have changed to an alpha-numeric input by using N$ in place of N. We have set N$ to 'null' in line 135. The new line 145 converts the alphanumeric variable N$ into a numeric variable N. In the version of BASIC used by many of the microcomputers, if N$ does not contain a numeric value, the variable N is set to zero and the test provided by the error sub-routine in line 1000 will fail and D will be returned equal to 1. If your version of BASIC produces an error when the string contains any non-numeric value, it is necessary to provide protection against this as well. In order to do this we must test that the input value is in fact a number before line 145 is reached. One way of doing this would be:

```
135  N$=" "
136  N=0
140  INPUT "(1,2 OR 3)";N$
141  IF N$<"1" OR N$ > "9" THEN 160
145  N=VAL(N$)
150  D=3
160  GOSUB 1000
165  IF D=1 THEN 135
```

This test takes place in line 141 which directs that if the input value is not a number, we go straight to the error sub-routine and as the value of N was set to zero in line 136, the test will fail.

These additional tests which have been introduced would normally be incorporated in the error sub-routine, otherwise they would have to be repeated for each question. If your machine is fitted with a bell or bleeper, it is quite useful to cause this to operate whenever an error message is printed. This is quite easily done by adding '; CHR$(7)' to the PRINT statement in line 1000 of programs 4(a) or 4(b).

Systems

Chapter 9 described how the simple search program 4(a) could be adapted so that the data could be stored on tape instead of including it in the program in the form of DATA statements. To do this we required a number of separate programs which produced a complete 'system'. If the memory capacity of your machine is large enough, all these programs could have been combined into one but this can lead to a much more complex program which is more difficult to debug or revise at a later date. Another problem is that if a sort program forms part of the system, a large amount of memory is usually required for the storage of the sort keys. The available memory for program text is therefore limited. If the solution that you choose in answer to a particular problem results in there being a number of programs required to make up a complete system, then a 'menu drive' should be considered. When this method is used the operator does not need to know the names which have been given to each individual program. A system of menu driven programs is more suitable for disk than for tape unless it is possible to rewind the tape under program control. A tape or disk which is to use this method will have one program called "START" or some other suitable name. If a cassette tape is to be used, this will be the first program on the tape, so that typing in LOAD will automatically load the START program. In the case of a disk, the name will have to be included in the LOAD statement unless your machine has a 'default' program name. In this case the default program name (if not START) should be used in place of START. The purpose of the START program is to display a menu on the screen which shows the purpose of each program which is contained in the system. The program required is then chosen by answering a suitable question and the system loads it from the disk or tape. The operator need have no knowledge of the names of the actual program being loaded because this information is contained within the "START" program. On completion of any of the programs in the system the option can be given to run the same program again or to return to the menu.

On some computers it is possible to expand the "START" program so that it will check that the hardware configuration is suitable for the particular system and, if more than one disk is needed for that system, whether the correct disk is present in the disk drive.

Program chaining

This has introduced a new principle, that of program chaining. It is possible to include a statement within a program which will automatically LOAD a new program and RUN it. The program which was previously in the memory of the machine is automatically cleared by this statement. For example, on the Commodore PET:

1500 LOAD "PROG 2"

will search for and load a program called "PROG 2" from the cassette tape. The existing program in memory will be cleared and the new program will be run.

As it is not possible to name a program which is stored on cassette tape, if the machine is an Apple the line would be written:

1500 LOAD

The required program must then be the next program on the tape.

In such cases, it is often desirable for the values of certain variables to be retained so that they can be used in the new program. If this is to be done, the variables must be declared as COMmon in the original program. To do this the COM statement is used. The values of any variables which are included in a COM statement are not affected when the new program is loaded — all other variables are cleared. This facility is not available on all microcomputers and one alternative is to store the required variables in a data file and then arrange for the new program to read the values from this file.

Program layout

When writing a program it is good practice to keep to a standard routine as far as possible. A program should normally contain its name together with the date of the current revision in its first line. This practice has not been followed in the programs included in the chapters, because in these situations it was considered unnecessary. It follows that whenever a revision is made to a program, the date contained in this line should be changed. Any program listing can then easily be identified as to whether it is the latest version or not.

Any variables which require dimensions should follow the identification line and then the main body of the program. All sub-routines and DATA lines (if any) should appear at the end. If similar sub-routines are used in several of your programs, it is a good plan to use the same line numbers in each program.

After some experience you may well find that the same routines are regularly required by your programs. Typical examples are input routines and routines for the revision of input. In this case it is a good plan to spend some time perfecting these sub-routines and storing them on disk as separate programs. These can then be incorporated in any of your programs which need them without having to retype them.

In conclusion it must be emphasised that we consider it essential that you learn the possibilities of your own machine and adapt these to your particular problems. It is not the kind of thing which can be absorbed overnight and will in fact take a few years of practice before you will feel reasonably confident that you know your machine. Perseverance is essential and it is hoped that this book will help to set a learner on the right track.

SECTION III
Starter Pack Programs

CHAPTER 11
Starter packs

One of the best ways of learning how to program a computer is to take an existing program and play with it. The process of keying the instructions into the machine helps to reinforce any knowledge gained already and hopefully a confidence in technique begins to develop. Almost any program that has been written has the capacity for improvement and refining an existing routine will again assist in developing confidence.

For this reason we have included a number of 'starter packs' of various length which illustrate several aspects of building appraisal. This should provide the reader with a head start for more ambitious programs that he may wish to develop.

The purpose of these programs is not to provide a commercial package but instead to provide an educational medium for learning how to use the machine. Most of the BASIC instructions used will be found in section II. The listings as reproduced have been obtained from a Commodore PET 4032 machine and this will differ very slightly from say the Apple, Tandy or other microcomputers. To be absolutely safe you should go back to the manual of your own machine, when a problem arises, to check that the BASIC is compatible. However to assist those with machines other than the PET, table 11.1 gives a list of those instructions included in the programs which are likely to be different between the various machines together with a translation for the Apple, Tandy, ZX81 and Superbrain (CP/M BASIC).

As these programs are essentially for educational purposes a number of simplifications have been made to make them understandable. Short cuts in technique are included with the descriptive matter of each pack or are obvious when running the program. More general simplifications are made here and include:

1. Most of the programs have been written with one BASIC statement on each line for ease of comprehension. Some machines will only allow one statement in any eventuality. If your machine is one of this type then you will need to separate the lines which have more than one statement into two or more, e.g.

   10 PRINT "MICRO" : PRINT " " : A = B * K

Table 11.1 Major differences in BASIC commands between leading brands of micros.

All the 'starter pack' programs in this book have been written to work as listed on the Commodore PET series of microcomputers (2001, 3000, 4000 and 8000 series). A minimum of 16K user memory store is necessary for the larger programs such as heat loss, developer's budget and cost plan. As already mentioned these starter packs have been so written as to facilitate easy conversion to other machines, and so are not necessarily written to run efficiently with regard to speed or length of coding. The major difference between machines is usually the means of clearing or homing the cursor on the screen; although these commands are not absolutely necessary they do improve presentation. There now follows a table of the screen control commands for some popular microcomputers which will need to be altered when the programs are being keyed in. The PET/VIC column shows the commands as they appear in the listings. The other columns show the equivalent command on other machines.

Make Function	PET/VIC	Apple II Applesoft	Integer	Tandy II	ZX81	Superbrain CP/M BASIC 2.2
Clear screen (cursor to top left position)	CHR$ (147)	HOME	CALL-936	CLS	CLS	CHR$ (12)
Home screen (as clear screen but all text remains on screen)	CHR$ (19)	VTAB 1: HTAB 1	N/A	CHR$ (28)	N/A but PRINT AT can be used	CHR$ (1)
GET a character (input a character from keyboard without displaying it)	GET ⟨String⟩ e.g. GET A$	GET ⟨String⟩ e.g. GET A$	N/A	⟨String⟩ =INKEY$	⟨String⟩ =INKEY$	⟨String⟩ =INPUT$(1)

In addition the following commands should be altered for Apple II and ZX81 users. (Note: the limited memory capacity of the ZX81, even with the 16K RAM pack, means that the longer programs are unlikely to be accepted in full.)

Apple II, ZX81 users
Statements such as

IF ⟨condition⟩ THEN ⟨Line No⟩

Table 11.1 Contd.

must be replaced by:

IF 〈condition〉 THEN GOTO 〈Line No〉

(THEN is optional on APPLE II)

ZX81 users
Statements such as

ON A GOTO 100, 200, 300

must be replaced by:

GOTO 100 * A

This can be any mathematical function.

---

would need to be separated into three lines, one for each statement separated by a colon.

```
10 PRINT "MICRO"
11 PRINT " "
12 A = B * K
```

2. Because of the length of the listings and the need to simplify the program it was decided to reduce the number of input checks that would normally be made. For a commercial package these checks would be essential.
3. The use of disks or printer has not been included as the manner of handling these peripherals between machines is extremely varied. However development of the programs will almost certainly require disk operation.
4. Simple BASIC has been used wherever possible and this should be common to nearly all machines which allow the use of BASIC language. However most modern machines have extra functions which allow more powerful routines to be written. These vary from use of the cursor to improved mathematical and formatting functions. When you are familiar with the common BASIC statements it is worth learning these new commands in order to make your program more efficient. In order to make our programs as straightforward as possible for the widest number of machines most of these have been excluded.
5. The formatting of the screen displays has been kept to a minimum as this is an area which is largely a matter of personal taste. The programs as written use a 40 × 25 screen but a number of screen widths and depths are available and this should not be considered standard. Alteration to screen displays will largely involve the PRINT and INPUT instructions, so attention should

be given to these in the first instance. There is no doubt that good formatting goes a long way to enhancing the pleasure of using the machine and even improving its credibility.

It is hoped that readers will enjoy the development of these programs and that they will be of practical benefit. If they do not appear to work in the first instance please check the keying in of instructions. The programs as listed have been checked several times and although it is still possible for errors to be included it is more likely that these will have occurred during the keyboard input.

S-curve cash flow projection

Introduction

This program forecasts the cash flow of a project using the Department of Health and Social Security formula (as described in the quarterly supplement to *The Chartered Surveyor*, 'Building and Quantity Surveying Quarterly', Volume 5, No. 3, spring 1978) for projects up to four million pounds. Actual valuations may also be input to allow comparison between forecast and actual progress.

Aim and objective

The aim is to show how a graph can be displayed using the simplest of graphics and yet give a clearer indication of the cash flow of the project than is possible from a table of figures. The program produces an S-curve on the screen against which actual valuations may be superimposed to enable the project programmer to see how the project is progressing. The program can also be used on several projects to build up a cash flow prediction for the financial year ahead.

Scope of program

Parameters

  (i) The program handles projects up to a four million pounds contract value.

 (ii) The maximum number of months that can be displayed at one time is twenty-one, but this can be altered if the check in line 205 is changed.

(iii) The program runs on forty column displays but may be tailored to individual machines if line 110 is altered.

(iv) The formula used to predict the cash flow is:

$$Y = S\,[X+CX^2-CX - \frac{1}{K}(6X^3 - 9X^2+3X)]$$

where Y is the cumulative monthly valuation (excluding retention), X is the month in which expenditure Y occurs divided by the contract period (i.e. proportion of contract complete), S is the contract sum, and C and K are constants depending on contract value.

Machine requirements
   (i) Central processing unit/keyboard
  (ii) Screen
 (iii) Cassette tape deck and blank tape or disk drive and disk for storing the program.

Data required
   (i) Contract sum (less retention) and contract period (months).
  (ii) Any actual valuations already paid.

Possible improvements

1. Plot S-curve using a printer to make hard copy (as with all starter packs, display is always to screen owing to the variety of printers available). This printout could include actual dates of valuations as most printers have more characters to a line than the screen width.
   Extra programming necessary to do this:
   (a) Input contract start (e.g. JAN 82).
   (b) Have months stored in a one-dimensional array of twelve elements, i.e. M$(1) = "JAN", M$(2) = "FEB" etc. Using the contract start date the month number can be identified and incremented on each successive valuation causing the next month's name to be printed out using the array, i.e.

   Month no.
   3   M$(3) = "MAR"
   4   M$(4) = "APR"

   A check must be made for the end of year on each iteration to reset the month number to one [M$(1) = "JAN"] and increment the year, if the last output month was twelve, i.e. M$(12) = "DEC".
   (c) Print the array variables for the months at the appropriate time.

2. The ability to store actual valuations on tape or disk. This would enable each actual valuation to be stored as it is entered so that all the office's projects may be updated on a monthly basis.
   Extra programming required:
   (a) Project identifier — such as a project number — to produce a directory of projects held on disk.
   (b) Menu to select required option, i.e. directory, new project, existing project or scratch (delete) project.

3. If the program is to be used as a cash flow estimating program

the facility to vary a single input parameter should be made available, so that a new set of estimated valuations may be found. In conjunction with an input contract start date the valuations for each month of a financial year may be totalled for all the office's projects to produce an estimated cash flow for the year.

4. Another possible extension is to take the actual valuations and conduct a regression analysis, to forecast when a project is likely to end, if the valuations were to continue on their present trend.

NB The program at present takes no account of any retention factor, as this may vary from contract to contract.

There follow the flow chart (fig. 11.1), the program listing, and the major screen displays (fig. 11.2) for 'S-curve'.

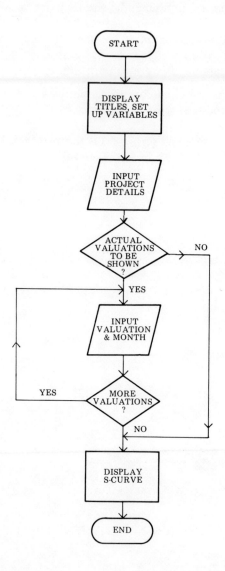

Fig. 11.1 Flow chart for 'S-curve'.

Program listing
S-curve


```
100 REM**************S-CURVE PROGRAM STARTER PK 15/06/82**************
105 REM*COPYRIGHT P.BRANDON,G.MOORE & P.MAIN WITH ACKNOWLEDGEMENT TO DHSS
110 LL=40
115 DIM A(21),T(8),C(8),K(8):REM                          <DIMENSION ARRAYS
120 PRINT CHR$(147);:REM                                  <CLEARS SCREEN!!!!!
125 PRINT"       DHSS EXPENDITURE FORECASTING":PRINT
130 PRINT"THIS PROGRAM ESTIMATES THE CUMULATIVE"
135 PRINT"MONTHLY VALUATIONS(EXCLUDING FLUCTUATIONS)"
140 PRINT"OF BUILDING PROJECTS FROM 10000 TO"
145 PRINT"4000000 POUNDS USING THE FORMULA:-"
150 PRINT"  V=CS(X+CX↑2-CX-(6X↑3-9X↑2+3X)/K)":PRINT:PRINT


155 REM************** INPUT CONTRACT DETAILS ***************
160 INPUT "WHAT IS THE CONTRACT SUM";CS
165 IF CS>5E6 OR CS<0 THEN PRINT"ERROR":GOTO 160
170 PRINT


175 REM**************      SELECTS CONSTANTS      **************
180 PRINT
185 FOR I=1 TO 8
190 READ T(I),C(I),K(I)
195 IF CS>T(I) THEN NEXT I
200 INPUT "WHAT IS THE CONTRACT LENGTH(MONTHS)";CL
205 IF CL>21 OR CL<0 THEN PRINT"ERROR":GOTO 200:REM     <MUST BE <=21 MONTHS


210 REM************** INPUT OF ANY VALUATIONS **************
215 PRINT CHR$(147);:REM                                  <CLEARS SCREEN!!!!!
220 PRINT"ACTUAL VALUATIONS (Y OR N)"
225 GET A$:IF A$<>"Y" AND A$<>"N" THEN 225
230 IF A$="N" THEN 275
235 PRINT:PRINT"TO FINISH TYPE '0'":PRINT
240 INPUT"MONTH";M
245 IF M=0 THEN 275
250 IF M>CL THEN PRINT"ERROR":GOTO 240
255 INPUT"VALUATION";A(M)
260 IF A(M)<0 OR A(M)>CS THEN PRINT"ERROR":GOTO 255
265 PRINT:GOTO 240


270 REM************** BEGIN SCREEN DISPLAY **************
275 PRINT CHR$(147);:REM                                  <CLEARS SCREEN!!!!!
280 PRINT"   CUMULATIVE MONTHLY VALUATION"
285 PRINT" 0  ---------------------------"


290 REM************** CALCS MONTHLY VALUATION **************
295 FOR J=1 TO CL
300  PRINTJ;TAB(4);
305  X=J/CL
310 V=CS*(X+C(I)*X↑2-C(I)*X-(6*X↑3-9*X↑2+3*X)/K(I)):REM  <DHSS FORMULA
```

Program listing
S-curve Contd.

```
315  L=INT((LL-15)*V/CS+.5):REM                          <FINDS BAR LENGTH
320   IF A(J)=0 THEN PT$="*":PS$="-":LO=0:HI=L:GOTO 385
325  H=INT((LL-15)*A(J)/CS+.5)
330  HI=L:LO=H:PT$="X":IF H>L THEN HI=H:LO=L:PT$="*"
335   IF LO<2 THEN 360


340  REM**************       POSITIONS CURSOR    **************
345   FOR X=1 TO LO-1:REM                              <DRAW LINE TO 1ST PT.
350   PRINT"-";
355   NEXT X
360   PRINT PT$;
365   IF PT$="*" THEN PT$="X":PS$=" ":GOTO 375
370  PT$="*":PS$="-"
375   IF H=L THEN 415
380   IF LO+1=HI THEN 405
385   IF HI<2 THEN 405
390   FOR X=LO+1 TO HI-1:REM                           <DRAW LINE TO 2ND PT.
395   PRINT PS$;
400   NEXT X
405  PRINT PT$;


410  REM************** PRINTS POINT & VALUATION**************
415  PRINT TAB((LL-2)-LEN(STR$(INT(V))));INT(V)
420  NEXT J
425  GET A$:IF A$="" THEN 425


430  REM************** DATA FOR CONSTANT SELECT**************
435  DATA 30000,-.409,7.018,75000,-.36,5,120000,-.24,4.932,300000,-.2,4.058
440  DATA 1200000,-.074,3.2,2000000,.01,3.98,3000000,.11,3.98,4000000,.159,3.78
9999 END
```

```
      DHSS EXPENDITURE FORECASTING

THIS PROGRAM ESTIMATES THE CUMULATIVE
MONTHLY VALUATIONS(EXCLUDING FLUCTUATIONS)
OF BUILDING PROJECTS FROM 10000 TO
4000000 POUNDS USING THE FORMULA:-
  V=CS(X+CX↑2-CX-(6X↑3-9X↑2+3X)/K)


WHAT IS THE CONTRACT SUM? 1000000


WHAT IS THE CONTRACT LENGTH(MONTHS)? 20
```

Screen 1

Explanation of program giving parameters and formula.

Project details:
Input contract sum between zero and £4 million.
Input contract length not exceeding 21 months.

```
ACTUAL VALUATIONS (Y OR N)

TO FINISH TYPE '0'

MONTH? 1
VALUATION? 20000

MONTH? 2
VALUATION? 45000

MONTH? 3
VALUATION? 90000

MONTH? 4
VALUATION? 120000

MONTH? 99
ERROR
MONTH? 0
```

Screen 2

This provides the opportunity to superimpose any actual valuations paid on to the predicted cash flow curve.

```
      CUMULATIVE MONTHLY VALUATION
 0  ---------------------------
 1  X                                   13436
 2  X                                   39160
 3  X*                                  75763
 4  -X*                                121840
 5  ---*                               175984
 6  -----*                             236790
 7  -------*                           302850
 8  --------*                          372760
 9  ----------*                        445111
10  ------------*                      518500
11  --------------*                    591518
12  ----------------*                  662760
13  ------------------*                730819
14  --------------------*              794290
15  ---------------------*             851765
16  -----------------------*           901840
17  ------------------------*          943106
18  ------------------------*          974160
19  -------------------------*         993593
20  -------------------------*        1000000
```

Screen 3

Display of curve:
* are the predicted valuations
X are the actual valuations.

Figures down the right hand side are the predicted valuations from the formula (not including retention).

Fig. 11.2 Major screen displays for 'S-curve'.

Reinforced concrete beam design

Introduction

This program is based on the British Standard Code of Practice CP 114 and is intended to design a simply supported beam in reinforced concrete, producing the required beam size and the cross-sectional area of the tensile reinforcement.

Aim and objective

The aim is to provide a simple program which can be used by a designer to obtain the approximate sizes of any simply supported reinforced concrete beam in a proposed building.

Scope of program

The beam breadth is taken to be equal to SPAN/30 and the depth of the beam required for this breadth is calculated taking into consideration its own weight. The minimum depth is taken as SPAN/20 (in accordance with CP 114). The program produces the required cross-sectional area of steel and also checks for shear. No details of shear reinforcement are given; the program simply states whether nominal or full shear reinforcement is required.

Machine requirements
 (i) Central processing unit/keyboard
(ii) Screen
(iii) Cassette tape deck and blank tape or disk drive and disk for storing the program.

Data requirements
The operator is first required to feed in the permissible stresses for the concrete and steel to be used. Following this, the span and the uniformly distributed load need to be entered. If the choice at the end of the program is to design more beams the permissible stresses are assumed to be the same as those entered on the first run. If it is desired to change these, the program must be run again from the beginning.

Possible alterations and development

(i)   Hard copy of results and input information.

(ii)   Allow for different breadths to be tried.

(iii) Adapt the program to calculate reinforced concrete slabs.

(iv)   Include as data, the standard bar sizes and extend the program to calculate the number and size of bars from the cross-sectional area.

(v)   Now that the bar sizes are available, the size and spacing of stirrups for shear reinforcement can be included.

(vi)   Arrange to produce the quantities of concrete and steel.

(vii) For those with a sufficient knowledge of structures the program can be extended to cater for other than simply supported beams, for example cantilever, continuous etc. If this is done it is probably better to write separate programs for the calculation of bending moment and shear force for the various cases and then chain these programs to the reinforced concrete design program. It is then possible to write a similar design program for steel beams and eventually produce a complete design system.

There follow the flow chart (fig. 11.3), the program listing, and the major screen displays (fig. 11.4) for 'Reinforced concrete beam design'.

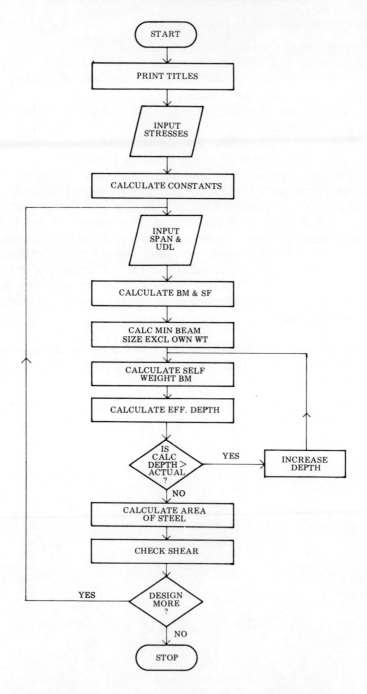

Fig. 11.3 Flow chart for 'Reinforced concrete beam design'.

Program listing
Reinforced concrete beam design

```
100 REM*************     BEAM CALCULATION STARTER PK 3/6/81      *************


105 REM*************  (C) COPYRIGHT BRANDON, MOORE & MAIN.   **************
110 PRINT CHR$(147);:REM                                    <CLEARS SCREEN!!!!
120 PRINT"THIS PROGRAM IS TO DESIGN A SIMPLY"
130 PRINT"SUPPORTED BEAM IN REINFORCED CONCRETE"
140 PRINT
150 PRINT"ASSUMPTIONS:-"
160 PRINT
170 PRINT
180 PRINT"MODULAR RATIO=15"
190 PRINT
200 PRINT"DENSITY OF CONCRETE=2500KG/CU M"
210 PRINT
220 PRINT"BEAM DEPTH=EFF DEPTH+38MM"
230 PRINT
240 PRINT"BEAM BREADTH NOT LESS THAN"
250 PRINT"SPAN/30(MIN 150MM)"
260 PRINT
270 PRINT"BEAM DEPTH NOT LESS THAN"
280 PRINT"SPAN/20(MIN 150MM)"
290 PRINT
300 PRINT"TOUCH 'RETURN' TO CONTINUE"
310 GET X$:IF X$<>CHR$(13) THEN 310
320 PRINT CHR$(147);:REM                                    <CLEARS SCREEN!!!!


330 REM*************          INPUT STRESS DATA         **************
340 INPUT"PERMISSIBLE BENDING STRESS IN CONCRETE  (N/SQ MM)";P0:PRINT
350 INPUT"PERMISSIBLE SHEAR STRESS IN CONCRETE    (N/SQ MM)";P1:PRINT
360 INPUT"PERMISSIBLE STRESS IN STEEL             (N/SQ MM)";P:PRINT


370 REM*************          CALCULATE CONSTANTS         **************
380 N1=1/(1+P/15/P0)
390 A1=1-N1/3
400 Q=P0/2*A1*N1


410 REM**********RESET BENDING MOMENT DUE TO  BEAM WEIGHT TO ZERO**********
420 M1=0


430 REM*************          INPUT BEAM DATA         **************
440 INPUT"WHAT IS THE SPAN OF THE BEAM(M)";L:PRINT
450 INPUT"WHAT IS THE U.D.LOAD(KN/M)      ";W:PRINT
460 PRINT:PRINT


470 REM************* CALCULATE BENDING MOMENT AND SHEAR FORCE **************
480 M=W*L↑2/8
490 S=W*L/2
```

Program listing
Reinforced concrete beam design Contd.

```
500 REM*************         CALCULATE MIN BREADTH           **************
510 B=INT(L/30*100+.9)/100
520 IF B>.15 THEN 540
530 B=.15


540 REM*************         CALCULATE MIN DEPTH             **************
550 D=INT(L/20*100+.9)/100
560 IF D>.15 THEN 580
570 D=.15


580 REM*********CALC BENDING MOMENT AND SHEAR FORCE DUE TO BEAM WT**********
590 M1=B*D*2.5*9.81*L↑2/8
600 S1=B*D*2.5*9.81*L/2


610 REM*************         CALCULATE EFF DEPTH             **************
620 D1=SQR((M+M1)*1E3/Q/B)
630 IF D1>112 THEN 650
640 D1=112
650 IF D1+38>L/20*1E3 THEN 670
660 D1=L/20*1E3-38
670 D2=INT((D1+38+9)/10)*10
680 IF D2<=D*1E3 THEN 720
690 D=INT((D2+10)*100+.9)/1E5


700 REM*************RECALCULATE BENDING MOMENT  DUE TO BEAM WT**************
710 GOTO 590
720 D1=D2-38


730 REM*************        CALCULATE AREA OF STEEL          **************
740 A=INT((M+M1)*1E6/P/A1/D1+.9)


750 REM*************                  CHECK SHEAR            **************
760 Q1=(S+S1)/B/A1/D1
770 PRINT"BEAM BREADTH =";B*1E3;"MM"
780 PRINT"BEAM DEPTH   =";D2;"MM"
790 PRINT"AREA OF STEEL=";A;"SQ MM"
800 IF Q1>P1 THEN 830
810 PRINT"NOMINAL SHEAR REINFORCEMENT REQUIRED"
820 GOTO 840
830 PRINT"SHEAR REINFORCEMENT REQUIRED"
840 PRINT
850 PRINT"ARE MORE BEAMS TO BE DESIGNED? (Y/N)"
860 GET Q$:IF Q$<>"Y" AND Q$<>"N" THEN 860
870 IF Q$="N" THEN 9999
880 PRINT CHR$(147);:REM                          <CLEARS SCREEN!!!!
890 GOTO 420
9999 END
```

```
THIS PROGRAM IS TO DESIGN A SIMPLY
SUPPORTED BEAM IN REINFORCED CONCRETE

ASSUMPTIONS:-


MODULAR RATIO=15

DENSITY OF CONCRETE=2500KG/CU M

BEAM DEPTH=EFF DEPTH+38MM

BEAM BREADTH NOT LESS THAN
SPAN/30(MIN 150MM)

BEAM DEPTH NOT LESS THAN
SPAN/20(MIN 150MM)

TOUCH 'RETURN' TO CONTINUE
```

Screen 1

States assumption for calculation.

```
PERMISSIBLE BENDING STRESS IN CONCRETE  (N/SQ MM)? 7

PERMISSIBLE SHEAR STRESS IN CONCRETE    (N/SQ MM)? 0.7

PERMISSIBLE STRESS IN STEEL             (N/SQ MM)? 140

WHAT IS THE SPAN OF THE BEAM(M)? 6

WHAT IS THE U.D.LOAD(KN/M)      ? 10


BEAM BREADTH = 200 MM
BEAM DEPTH   = 510 MM
AREA OF STEEL= 994 SQ MM
NOMINAL SHEAR REINFORCEMENT REQUIRED

ARE MORE BEAMS TO BE DESIGNED? (Y/N)
```

Screen 2

Requests design parameters and outputs beam breadth and depth.

Fig. 11.4 Major screen displays for 'Reinforced concrete beam design'.

Cost analysis and cost plan

Introduction

These two programs allow the numerical information of an elemental cost analysis to be placed on to a tape file and for that information to be retrieved from tape and adjusted for a new project.

Aim and objective

The aim is to demonstrate how a tape file can be created for elemental information and to show how the machine can be used at the very early stages of budgetary control to provide a continuous review of the current state of the cost plan. Adjustments are allowed for single elements at the end of an initial program run in order that the budget targets can be 'fine tuned'.

Scope of program

Certain simplifications have been made for the sake of clarity including:

(i) Specification details have not been included.
(ii) Adjustment for specification is made by a percentage addition to or subtraction from the figure adjusted for quality.
(iii) Printing of a hardcopy has not been included.
(iv) Input checks have been kept to a minimum.
(v) Preliminaries can be included as either a percentage of the remainder of the contract or adjusted in a similar way to the specification in the other elements (i.e. percentage up or down as the analysis figure). Preliminaries should always be the last element to be entered from the analysis.
(vi) The file uses the name of the project, e.g. 'Factory 1', as the means of finding the element on tape. Do not have two projects with the same name.

Machine requirements
(i) Central processing unit/keyboard
(ii) Screen
(iii) Cassette tape deck for recording analysis and a blank tape.

Data requirements

The operator will need to have before him for the program of COST ANALYSIS:

(i)   A copy of an elemental cost analysis to be placed on file (in any format, except 'preliminaries' must be the last element).

(ii) A unique name for the analysis.

For the program of COST PLAN he requires:

(i)   The tape containing the relevant historic analysis.

(ii) The quantities of the new project required for inputting the element unit quantities.

(iii) A knowledge of the location of the project.

(iv) An indication of the new specification cost in relation to the cost analysis specification.

(v) The current tender-based index.

Possible alterations and development

The program represents a very basic package but it can provide the foundation for several future developments, e.g.

(i)   Hardcopy of results and intermediate calculations.

(ii) Output and input of analyses from disk instead of tape.

(iii) Inclusion of a specification description in the analysis and cost plan.

(iv) Other programs which will access the data file of analyses and analyse the information held for other purposes e.g. forming distributions of elemental cost. It will be essential to store the information on disks if this is to be attempted.

(v) If the results of the budget are output to file then these results can be called up for use with a program using more detailed measurements as design develops. The budget can then be used as a point of reference for further monitoring.

Note: the programs are written for a Commodore PET. The few instructions which may differ for other machines are included in table 11.1.

There follow the flow chart (fig. 11.5), the program listing, and the major screen displays (fig. 11.6) for 'Cost analysis', and the same for 'Cost plan' (fig. 11.7, listing, and fig. 11.8).

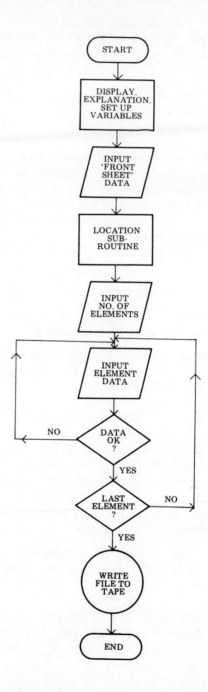

Fig. 11.5 Flow chart for 'Cost analysis'.

Program listing
Cost analysis

```
100 REM************* COST ANALYSIS SETUP PROGRAM **************


105 REM******(C) COPYRIGHT BRANDON,MOORE & MAIN. 18-05-82*******
110 DIM A$(40),FS$(10):REM                                  <DIMENSION ARRAYS
115 PRINT CHR$(147);:REM                                     <CLEARS SCREEN!!!!
120 PRINT"        $$$$$ $$$$$ $$$$$ $$$$$"
125 PRINT"        $$    $$ $ $$       $"
130 PRINT"        $$    $$ $ $$$$$    $"
135 PRINT"        $$    $$ $     $    $"
140 PRINT"        $$$$$ $$$$$ $$$$$    $":PRINT
145 PRINT"                ANALYSIS":PRINT
150 PRINT"  THIS PROGRAM WRITES COST ANALYSIS":PRINT
155 PRINT"  DATA TO FILE FOR USE WITH THE COST":PRINT
160 PRINT"  COST PLANNING EXAMPLE PROGRAM.":PRINT
165 PRINT"  PRESS 'RETURN' TO CONTINUE."
170 GET ZZ$:IF ZZ$<>CHR$(13) THEN 170


175 REM*************    FRONT SHEET DATA       **************
180 PRINT CHR$(147);:REM                                    <CLEARS SCREEN!!!!
185 PRINT"LOAD TAPE & WIND TO END OF LAST DATA":PRINT
190 PRINT:PRINT"INPUT THE FOLLOWING DATA.(MAX 16 CHARS)":PRINT
195 DATA "ANALYSIS TITLE   :","BUILDING TYPE   :","LOCATION        :"
200 DATA "TENDER DATE      :","TENDER INDEX    :","CONTRACT        :"
205 DATA "FLOOR AREA       :","NO. OF STOREYS  :"
210 FOR X=1 TO 8
215 READ QU$
220 PRINT QU$;
225 INPUT FS$(X)
230 IF LEN(FS$(X))>16THEN PRINT"ERROR,TOO LONG":GOTO 220
235 NEXT X
240 GOSUB 400:REM                                           <LOCN SUBROUTINE


245 REM************* INPUT OF ELEMENTAL BREAKDOWN*************
250 PRINT CHR$(147);:REM                                     <CLEARS SCREEN!!!!
255 PRINT"ENTRY OF ELEMENTAL DATA":PRINT
260 INPUT"HOW MANY ELEMENTS(40 MAX)";N
265 IF N<0 OR N>40 THEN PRINT"ERROR":GOTO 260
270 FOR K=1 TO N
275 PRINT CHR$(147);:REM                                     <CLEARS SCREEN!!!!
280 PRINT"  INPUT THE DATA FOR ELEMENT NO";K:PRINT
285 PRINT"  ELEMENT NAME       EUR    COST/M2"
290 PRINT" ---------------    ---.--   ---.--"
295 INPUT A$(K):PRINT
300 IF LEN(A$(K))<30 OR LEN(A$(K))>38 THEN 275
305 PRINT"DATA OK(Y/N)"
310 GET ZZ$:IF ZZ$<>"Y" AND ZZ$<>"N" THEN 310
315 IF ZZ$="Y" THEN 325
320 GOTO 275
325 K=K+1:IF K<=N THEN 275
```

Program listing
Cost analysis Contd.

```
330 REM************* WRITING DATA TO CASSETTE  **************
335 OPEN2,1,1,FS$(1):REM                                  *<OPEN CHANNEL
340 PRINT"DONT PANIC I'M WRITING YOUR DATA TO FILE":REM   *
345 FOR K=1 TO 9:REM                                      *
350 PRINT#2,FS$(K);CHR$(13);:REM                          *<WRT FRONT SHEET
355 NEXT K:REM                                            *
360 PRINT#2,N;CHR$(13);:REM                               *<WRT NO.ELEMENTS
365 FOR K=1 TO N:REM                                      *
370 PRINT#2,A$(K);CHR$(13);:REM                           *<WRT ELEM.DATA
375 NEXT K:REM                                            *
380 CLOSE 2:REM**********************************************<CLOSE CHANNEL
385 PRINT"ANALYSIS WRITTEN TO FILE-GOODBYE"
390 GOTO 9999


395 REM************** LOCATION SUBROUTINE  **************
400 PRINT CHR$(147);:REM                            <CLEARS SCREEN!!!!
405 PRINT
410 PRINT"    IN WHICH OF THE FOLLOWING"
415 PRINT"    -------------------------"
420 PRINT"  LOCATIONS IS YOUR SITE SITUATED"
425 PRINT"  -------------------------------"
430 PRINT"1. OUTER LONDON .. .. .. ..   1.00"
435 PRINT"2. INNER LONDON .. .. .. ..   1.08"
440 PRINT"3. LIVERPOOL.. .. .. .. ..    1.05"
445 PRINT"4. SOUTH EAST. .. .. .. ..    0.95"
450 PRINT"5. SCOTLAND & WALES .. .. ..  0.93"
455 PRINT"6. EAST OR WEST MIDLANDS . ..  0.92"
460 PRINT"7. NORTH WEST OR YORKS/HUMBER  0.89"
465 PRINT"8. EAST ANGLIA .. .. .. ..    0.91"
470 PRINT"9. NORTHERN OR SOUTHWEST . ..  0.88":PRINT
475 PRINT"ENTER WEIGHTING FACTOR BASED ON THE AJ"
480 PRINT"WEIGHTINGS ABOVE OR YOUR OWN JUDGEMENT":PRINT
485 INPUT"       WEIGHTING FACTOR =";X
490 IF X<0.5 OR X>1.5 THEN 500
495 FS$(9)=STR$(X):RETURN
500 PRINT"UNREASONABLE ENTRY-TRY AGAIN":GOTO 485
9999 END
```

If you wish to output the analysis information to disk then line 335 must be altered. For PET disks line 335 must be changed to:

335 OPEN 2, 8, 2, "O:" + FS$(1) + ",S, W"

```
LOAD TAPE & WIND TO END OF LAST DATA


INPUT THE FOLLOWING DATA.(MAX 16 CHARS)

ANALYSIS TITLE    :? EXAMPLE ANALYSIS
BUILDING TYPE     :? FACTORY
LOCATION          :? WARWICK
TENDER DATE       :? 21/02/81
TENDER INDEX      :? 150
CONTRACT          :? JCT STANDARD
FLOOR AREA        :? 3000
NO. OF STOREYS    :? 2
```

Screen 1

Requests the cost analysis 'front sheet' information. The input data will be used in 'Cost plan' for updating and adjustment.

```
        IN WHICH OF THE FOLLOWING
        -------------------------
    LOCATIONS IS YOUR SITE SITUATED
    -------------------------------
1.  OUTER LONDON ..  ..  ..  ..    1.00
2.  INNER LONDON ..  ..  ..  ..    1.08
3.  LIVERPOOL..  ..  ..  ..  ..    1.05
4.  SOUTH EAST.  ..  ..  ..  ..    0.95
5.  SCOTLAND & WALES ..  ..  ..    0.93
6.  EAST OR WEST MIDLANDS .  ..    0.92
7.  NORTH WEST OR YORKS/HUMBER     0.89
8.  EAST ANGLIA  ..  ..  ..  ..    0.91
9.  NORTHERN OR SOUTHWEST .  ..    0.88

ENTER WEIGHTING FACTOR BASED ON THE AJ
WEIGHTINGS ABOVE OR YOUR OWN JUDGEMENT

        WEIGHTING FACTOR =? .9
```

Screen 2

Requests the input of region for adjustment of location in 'Cost plan'.

These location factors are reproduced by kind permission of the *Architects' Journal* **and Messrs Davis, Belfield and Everest (chartered quantity surveyors)**

```
  INPUT THE DATA FOR ELEMENT NO 1

   ELEMENT NAME      EUR    COST/M2
   ------------      ---.--   ---.--
? SUBSTRUCTURE       17.50    16.80

DATA OK(Y/N)
```

Screen 3

**Requests the input of numerical information for each element. It is important to position the descriptions and decimal points in the right place.
After the final input it writes the data to tape file.**

Fig. 11.6 Major screen displays for 'Cost analysis'.

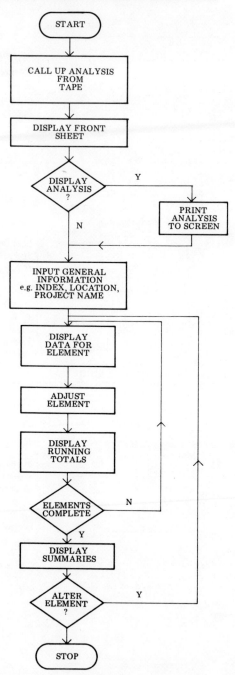

Fig. 11.7 Flow chart (simplified) for 'Cost plan'.

Program listing
Cost plan

```
1000 REM************** COST PLAN STARTER PACK 18-05-82 **************


1005 REM*************(C) COPYRIGHT BRANDON,MOORE & MAIN.**************
1010 DIM A$(40),B$(40),C$(40),D$(40),B(40),C(40),E(40,5),FS$(10)
1015 PRINT CHR$(147):PRINT:PRINT:REM                    <CLEARS SCREEN!!!!!


1020 REM*************** TITLES AND EXPLANATION OF PROG ***************
1025 PRINT"$$$$ $$$$ $$$$ $$$$ $$$$ $$    $$$$ $  $"
1030 PRINT"$$$$ $$$$ $$$$ $$$$ $$$$ $$    $$$$ $  $"
1035 PRINT"$$   $ $ $$    $$  $ $ $$   $  $ $$ $"
1040 PRINT"$$   $ $ $$$$  $$   $$$$ $$   $$$$ $ $$"
1045 PRINT"$$   $ $  $$   $$  $$   $$   $  $ $  $"
1050 PRINT"$$$$ $$$$ $$$$  $$   $$   $$$$ $  $ $  $":PRINT
1055 GOSUB 2575:REM                                     <PAUSE ROUTINE.
1060 PRINT CHR$(147):PRINT:PRINT:REM                    <CLEARS SCREEN!!!!!
1065 PRINT"THIS PROGRAM DEMONSTRATES THE USE OF"
1070 PRINT:PRINT"THE MICROCOMPUTER TO TAKE A COST"
1075 PRINT:PRINT"ANALYSIS AND UPDATE IT QUICKLY AND"
1080 PRINT:PRINT"ACCURATELY FOR ANOTHER PROJECT."
1085 PRINT:PRINT"QUANTITY,QUALITY PRICE AND LOCATION"
1090 PRINT:PRINT"ADJUSTMENTS CAN BE MADE AT THE"
1095 PRINT:PRINT"DISCRETION OF THE USER.":PRINT
1100 GOSUB 2575:REM                                     <PAUSE ROUTINE.
1105 PRINT CHR$(147):PRINT:PRINT:REM                    <CLEARS SCREEN!!!!!
1110 PRINT"ANSWER THE QUESTIONS AND THEN PRESS"
1115 PRINT"RETURN AFTER EACH ANSWER....":PRINT


1120 REM*************** INPUT OF ANALYSIS FROM TAPE  ***************
1125 INPUT"TITLE OF ANALYSIS    ";NAM$
1140 PRINT:PRINT"I AM ABOUT TO LOOK FOR YOUR ANALYSIS":PRINT
1145 PRINT"     CALLED ";NAM$


1150 REM*************** RETRIEVAL OF INFO FROM TAPE  ***************
1155 OPEN 2,1,0,NA$:REM                                 <OPEN TAPE CHANNEL
1160 PRINT:PRINT"DONT GO AWAY-I HAVE FOUND YOUR ANALYSIS":PRINT
1165 PRINT"     CALLED ";NAM$
1170 FOR K=1 TO 9
1175 INPUT#2,FS$(K):REM                                 <FRONT SHEET INFO
1180 NEXT K
1185 INPUT#2,N:REM                                      <NO. OF ELEMENTS
1190 FOR K=1 TO N
1195 INPUT#2,A$(K):REM                                  <ELEMENT INFO
1200 NEXT K
1205 CLOSE 2:REM                                        <CLOSE TAPE CHANNEL


1210 REM*************** SUBROUTINE FOR ANALYSIS LAYOUT ***************
1215 PRINT CHR$(147):PRINT:PRINT:REM                    <CLEARS SCREEN!!!!!
1220 PRINT"YOUR ANALYSIS HAS BEEN FOUND & ENTERED"
```

Program listing
Cost plan Contd.

```
1225 PRINT" !================================!"
1230 PRINT" !";TAB(10);FS$(1);TAB(34);"!"
1235 PRINT" !================================!"
1240 GOSUB 2575:REM                              <PAUSE ROUTINE.
1245 PRINT CHR$(147):REM                         <CLEARS SCREEN!!!!!
1250 PRINT" !--------------------------------!"
1255 PRINT" ! BUILDING TYPE:                 !"
1260 PRINT" !--------------------------------!"
1265 PRINT" ! LOCATION    :                  !"
1270 PRINT" !--------------------------------!"
1275 PRINT" ! TENDER DATE  :                 !"
1280 PRINT" !--------------------------------!"
1285 PRINT" ! TENDER INDEX :                 !"
1290 PRINT" !--------------------------------!"
1295 PRINT" ! CONTRACT    :                  !"
1300 PRINT" !--------------------------------!"
1305 PRINT" ! FLOOR AREA  :                  !"
1310 PRINT" !--------------------------------!"
1315 PRINT" ! NO OF STOREYS:                 !"
1320 PRINT" !--------------------------------!"
1325 PRINT CHR$(19):PRINT:REM                    <HOMES SCREEN !!!!!
1330 FOR K=2 TO 8
1335 PRINTTAB(17);FS$(K):PRINT
1340 NEXT K
1345 GOSUB 2575:REM                              <PAUSE ROUTINE.


1350 REM*************** FILL ARRAYS WITH ELEMENT FIELDS***************
1355 FOR K=1 TO N
1360 D$(K)=LEFT$(A$(K),16)
1365 B$(K)=MID$(A$(K),18,7):B(K)=VAL(B$(K))
1370 C$(K)=RIGHT$(A$(K),7):C(K)=VAL(C$(K))
1375 NEXT K
1380 FS(5)=VAL(FS$(5))


1385 REM***************     DISPLAY ELEMENTAL DATA    ***************
1390 PRINT CHR$(147):REM                         <CLEARS SCREEN!!!!!
1395 PRINT"DO YOU WISH TO SEE ELEMENTAL DATA(Y/N)"
1400 GET M$:IF M$<>"Y" AND M$<>"N" THEN 1400
1405 IF M$="N" THEN 1495
1410 PRINT CHR$(147);:REM                        <CLEARS SCREEN!!!!!
1415 PRINT"ELEMENT",,"  EUR","COST/M2"
1420 PRINT"------------------------------------"
1425 T=0:REM                                     <SET COUNT
1430 FOR K=1 TO 40
1435 PRINT D$(K),B$(K),C$(K)
1440 IF K>=N THEN 1490
1445 PRINT" "
1450 IF K<(T+9) THEN 1480
1455 GOSUB 2575:REM                              <PAUSE ROUTINE.
1460 T=T+9
1465 PRINT CHR$(147);:REM                        <CLEARS SCREEN!!!!!
1470 PRINT"ELEMENT",,"  EUR","COST/M2"
1475 PRINT"------------------------------------"
```

Program listing
Cost plan Contd.

```
1480 NEXT K


1485 REM***************          COST PLAN ROUTINE      ***************
1490 GOSUB 2575:REM                                      <PAUSE ROUTINE.
1495 PRINT CHR$(147):PRINT:PRINT:REM                     <CLEARS SCREEN!!!!!
1500 PRINT"COST PLAN COMPUTATION USING THE ANALYSIS":PRINT
1505 PRINT"THIS PART OF THE PROGRAM ALLOWS YOU TO":PRINT
1510 PRINT"ADJUST THE ANALYSIS DATA FOR CHANGES":PRINT
1515 PRINT"IN QUALITY, QUANTITY, PRICE & LOCATION.":PRINT
1520 PRINT"THE CALCULATION IS BASED UPON :-":PRINT
1525 PRINT" EUR(ANALYSIS)    *   EUQ(PROJECT)"
1530 PRINT"                     ------------"
1535 PRINT"                     GFA(PROJECT)":PRINT
1540 PRINT"'PRICE'   IS ADJUSTED USING THE INDEX":PRINT
1545 PRINT"'LOCATION' BY A WEIGHTING FACTOR":PRINT
1550 GOSUB 2575:REM                                      <PAUSE ROUTINE.
1555 PRINT CHR$(147):PRINT:PRINT:REM                     <CLEARS SCREEN!!!!!
1560 PRINT"  COST PLAN BASED UPON ";FS$(1):PRINT:PRINT
1565 PRINT"PLEASE GIVE THE FOLLOWING INFORMATION":PRINT
1570 PRINT"ON THE PROJECT TO BE COST PLANNED":PRINT
1575 PRINT:INPUT"GROSS FLOOR AREA OF PROJECT(M2)";GFA
1580 PRINT:INPUT"CURRENT INDEX VALUE       ";I1
1585 PRINT:INPUT"TENDER INDEX FORECAST      ";I2
1590 PRINT:INPUT"PROJECT NAME";P$
1595 PRINT:INPUT"TODAYS DATE ";TD$


1600 REM***************          LOCATION SUBROUTINE     ***************
1605 PRINT CHR$(147):PRINT:PRINT:REM                     <CLEARS SCREEN!!!!!
1610 PRINT"     IN WHICH OF THE FOLLOWING"
1615 PRINT"     ------------------------"
1620 PRINT"   LOCATIONS IS YOUR SITE SITUATED"
1625 PRINT"   ------------------------------":PRINT
1630 PRINT" 1-OUTER LONDON               1.00"
1635 PRINT" 2-INNER LONDON               1.08"
1640 PRINT" 3-LIVERPOOL                  1.05"
1645 PRINT" 4-SOUTH EAST                 0.95"
1650 PRINT" 5-SCOTLAND & WALES           0.93"
1655 PRINT" 6-EAST OR WEST MIDLANDS      0.92"
1660 PRINT" 7-NORTH WEST OR YORKS/HUMBER 0.89"
1665 PRINT" 8-EAST ANGLIA                0.91"
1670 PRINT" 9-NORTHERN OR SOUTHWEST      0.88":PRINT
1675 PRINT"ENTER WEIGHTING FACTOR BASED ON THE AJ"
1680 PRINT"WEIGHTINGS ABOVE OR YOUR OWN JUDGEMENT":PRINT
1685 INPUT"       WEIGHTING FACTOR =";X
1690 IF X<0.5 OR X>1.5 THEN 1700
1695 GOTO 1705
1700 PRINT"UNREASONABLE ENTRY-TRY AGAIN":GOTO 1685
1705 VL=VAL(FS$(9)):REM                                  <INDEX OF ANALYSIS
1710 V1=I1*(X/VL):REM                                    <ADJUSTMENT(CURRENT)
1715 V2=I2*(X/VL):REM                                    <ADUSTMENT(FUTURE)
1720 FOR K=1 TO 40
1725 PRINT CHR$(147):REM                                 <CLEARS SCREEN!!!!!
1730 IF K>N THEN 2110
```

Program listing
Cost plan Contd.

```
1735 REM***************     DISPLAY ELEMENT DATA      ***************
1740 PRINT"  COST PLAN BASED ON ";FS$(1):PRINT
1745 PRINT"ELEMENT NO ";K,D$(K):PRINT
1750 PRINT"              ANALYSIS DATA"
1755 PRINT"=========================================":PRINT
1760 PRINT"ELEMENT UNIT RATE=          ";:ZZ=VAL(B$(K)):GOSUB 2120
1765 E1=INT((B(K)*V1/FS(5))*100+0.5)/100
1770 PRINT:PRINT"EUR UPDATED TO TODAYS DATE=   ";:ZZ=E1:GOSUB 2120
1775 PRINT:PRINT"COST/M2 OF G F A =          ";:ZZ=VAL(C$(K)):GOSUB 2120
1780 E2=INT((C(K)*V1/FS(5))*100+0.5)/100
1785 PRINT:PRINT"COST/M2 @ TODAYS DATE=       ";:ZZ=E2:GOSUB 2120:PRINT
1790 PRINT"=========================================" 
1795 PRINT"              PROJECT DATA"
1800 IF K<N THEN 1895
1805 IF LEFT$(D$(K),6)="PRELIM" THEN 1815
1810 GOTO 1895
1815 PRINT:PRINT"DO YOU WANT TO:-1 ADJUST ABOVE RATE"
1820 PRINT:PRINT"              2 INCLUDE PRELIMS AS %"
1825 INPUT"TYPE 1 OR 2";M:IF M<1 OR M>2 THEN PRINT"ERROR":GOTO 1825
1830 IF M<2 THEN 1895


1835 REM***************     SUBROUTINE FOR PRELIMS     ***************
1840 INPUT"WHAT % DO YOU WISH TO ADD TO TOTAL";P
1845 IF P<0 OR P>100 THEN PRINT"ERROR":GOTO 1840
1850 E(K,1)=GFA:E(K,4)=T*(P/100.)
1855 E(K,5)=T1*(P/100)
1860 E(K,4)=INT(E(K,4)*100+0.5)/100
1865 E(K,5)=INT(E(K,5)*100+0.5)/100
1870 E(K,2)=E(K,4)
1875 E(K,3)=E(K,5)
1880 R1=K
1885 GOTO 1945


1890 REM***************     SPECIFICATION CHANGE      ***************
1895 PRINT:INPUT"ELEMENT UNIT QUANTITY(PROJECT)";E(K,1)
1900 PRINT:INPUT"SPECIFICATION CHANGE(% + OR -)";SC
1905 E(K,2)=(B(K)+(B(K)*SC/100))*V1/FS(5)
1910 E(K,2)=INT(E(K,2)*100+0.5)/100
1915 E(K,3)=(B(K)+(B(K)*SC/100))*V2/FS(5)
1920 E(K,3)=INT(E(K,3)*100+0.5)/100
1925 E(K,4)=E(K,2)*E(K,1)/GFA
1930 E(K,4)=INT(E(K,4)*100+0.5)/100
1935 E(K,5)=E(K,3)*E(K,1)/GFA
1940 E(K,5)=INT(E(K,5)*100+0.5)/100
1945 T=T+E(K,4):T1=T1+E(K,5)
1950 T1=INT((T1*100)+0.5)/100
1955 IF V>0 THEN 2100
1960 PRINT CHR$(147):REM                           <CLEARS SCREEN!!!!!
1965 PRINT"    SUMMARY OF ELEMENTAL COSTS":PRINT
1970 PRINT"              ELEMENT NO";K:PRINT
1975 PRINT"         ";D$(K)
```

Program listing
Cost plan Contd.

```
1980 PRINT"ELEMENT UNIT QUANTITY          ";:ZZ=E(K,1):GOSUB 2120
1985 PRINT"ELEMENTAL UNIT RATE(CURRENT) ";:ZZ=E(K,2):GOSUB 2120
1990 PRINT"ELEMENTAL UNIT RATE(TENDER)  ";:ZZ=E(K,3):GOSUB 2120:PRINT
1995 IF M<>2 THEN 2005
2000 PRINT"PRELIMINARIES INCLUDED AT ";P;" %":PRINT
2005 PRINT"ELEMENTAL COST/M2(CURRENT)   ";:ZZ=E(K,4):GOSUB 2120
2010 PRINT"ELEMENTAL COST/M2(TENDER)    ";:ZZ=E(K,5):GOSUB 2120:PRINT
2015 PRINT"TOTAL COST OF ELEMENT(TENDER)";:ZZ=E(K,5)*GFA:GOSUB 2120:PRINT
2020 PRINT"RUNN'G ELEMENTS TOT. COST/M2";:ZZ=T1:GOSUB 2120
2025 PRINT"RUNNING TOTAL OF ELEMENTS   ";:ZZ=(T1*GFA):GOSUB 2120:PRINT
2030 PRINT"ARE YOU HAPPY WITH THESE FIGURES(Y/N)"
2035 GET M$:IF M$<>"Y" AND M$<>"N" THEN 2035
2040 IF M$="Y" THEN 2070
2045 T=T-E(K,4):T1=T1-E(K,5)
2050 PRINT CHR$(147):REM                              <CLEARS SCREEN!!!!!
2055 PRINT"REPEAT OF ELEMENT NO";K;D$(K)
2060 GOTO 1750


2065 REM***************       BRANCHING ROUTINE      ***************
2070 IF R>0 THEN 2080
2075 GOTO 2100
2080 K=R
2085 T=T-E(K,4):T1=T1-E(K,5)
2090 V=1
2095 GOTO 1845
2100 IF Q>0 THEN 2375
2105 NEXT K
2110 GOTO 2165


2115 REM*************** SUBROUTINE TO POSITION RESULTS ***************
2120 TT=33:ZY=ZZ
2125 IF ZY>9.99 THEN ZY=ZY/10:TT=TT-1:GOTO 2125
2130 PRINT TAB(TT);INT(0.5+ZZ*100)/100
2135 RETURN
2140 TT=23:ZY=ZZ
2145 IF ZY>9.99 THEN ZY=ZY/10:TT=TT-1:GOTO 2145
2150 PRINT TAB(TT);INT(0.5+ZZ*100)/100;
2155 RETURN


2160 REM***************       SUMMARY ELEMENTAL COSTS   ***************
2165 PRINT CHR$(147):PRINT:REM                        <CLEARS SCREEN!!!!!
2170 PRINT"   WHICH OUTPUT DO YOU REQUIRE:-"
2175 PRINT:PRINT"   1. SUMMARY AT TODAYS DATE"
2180 PRINT:PRINT"   2. SUMMARY AT TENDER DATE":PRINT
2185 INPUT"TYPE APPROPRIATE NUMBER";Z:IF Z<1ORZ>2 THEN PRINT"ERROR":GOTO 2185
2190 IF Z=2 THEN 2200
2195 IN=I1:GOTO 2205
2200 IN=I2
2205 Y=Z+3:Z=Z+1
2210 PRINT CHR$(147);:REM                             <CLEARS SCREEN!!!!!
2215 PRINT"   ELEMENTAL COST SUMMARY":PRINT
2220 PRINT"ELEMENT",,"  EUR"," COST/M2"
```

Program listing
Cost plan Contd.

```
2225 PRINT"-------------------------------------"
2230 L=0
2235 FOR K=1 TO 40
2240 ZZ=E(K,Z):ZX=E(K,Y):PRINT D$(K),:GOSUB 2140:ZZ=ZX:GOSUB 2120
2245 IF K>=N THEN 2290
2250 PRINT""
2255 IF K<(L+9) THEN 2285
2260 GOSUB 2575:REM                              <PAUSE ROUTINE.
2265 L=L+9
2270 PRINT CHR$(147);:REM                        <CLEARS SCREEN!!!!!
2275 PRINT"ELEMENT",," EUR"," COST/M2"
2280 PRINT"-------------------------------------"
2285 NEXT K
2290 GOSUB 2575:REM                              <PAUSE ROUTINE.
2295 PRINT CHR$(147);:REM                        <CLEARS SCREEN!!!!!
2300 PRINT:PRINT"========================================="
2305 PRINT:PRINTP$;,,TD$
2310 PRINT:PRINT"FLOOR AREA";GFA,"INDEX";IN
2315 PRINT:PRINT"        COST/M2 SUMMARY"
2320 PRINT:PRINT"                         POUNDS"
2325 PRINT:PRINT"COST/M2(CURRENT PRICES)  ";:ZZ=T:GOSUB 2120
2330 PRINT:PRINT"TOTAL COST(DITTO)        ";:ZZ=(T*GFA):GOSUB 2120
2335 PRINT:PRINT"COST/M2(TENDER DATE)     ";:ZZ=T1:GOSUB 2120
2340 PRINT:PRINT"TOTAL COST(DITTO)        ";:ZZ=(T1*GFA):GOSUB 2120
2345 PRINT:PRINT"========================================="
2350 PRINT"DO YOU WANT ANOTHER SUMMARY(Y/N)"
2355 GET M$:IF M$<>"Y" AND M$<>"N" THEN 2355
2360 Z=0:Y=0
2365 IF M$="Y" THEN 2165


2370 REM***************    CHANGE ELEMENT ROUTINE    ***************
2375 PRINT CHR$(147):PRINT:PRINT:REM              <CLEARS SCREEN!!!!!
2380 PRINT"DO YOU WANT TO ALTER AN ELEMENT (Y/N)?"
2385 Q=0:V=0
2390 GET M$:IF M$<>"Y" AND M$<>"N" THEN 2390
2395 IF M$="Y" THEN 2405
2400 GOTO 2540
2405 PRINT CHR$(147);:REM                         <CLEARS SCREEN!!!!!
2410 PRINT"        ELEMENT LISTING"
2415 PRINT"   (NOTE ELEMENT NO TO BE CHANGED)":PRINT
2420 M=0
2425 PRINT"ELEMENT",,"COST/M2"," NO"
2430 PRINT"-------------------------------------"
2435 L=0
2440 FOR K=1 TO 40
2445 ZZ=E(K,5):PRINT D$(K),:GOSUB 2140:PRINT,K
2450 IF K>=N THEN 2495
2455 PRINT
2460 IF K<(L+9) THEN 2490
2465 GOSUB 2575:REM                               <PAUSE ROUTINE.
2470 L=L+9
2475 PRINT CHR$(147);:PRINT:PRINT:REM             <CLEARS SCREEN!!!!!
2480 PRINT"ELEMENT",,"COST/M2"," NO"
2485 PRINT"-------------------------------------"
```

Program listing
Cost plan Contd.

```
2490 NEXT K
2495 GOSUB 2575:REM                           <PAUSE ROUTINE.
2500 PRINT CHR$(147):PRINT:PRINT:REM          <CLEARS SCREEN!!!!!
2505 INPUT"ENTER NO OF ELEMENT TO BE CHANGED";K
2510 IF K<1 OR K>N THEN 2500
2515 Q=1
2520 R=R1
2525 T=T-E(K,4):T1=T1-E(K,5)
2530 GOTO 1725


2535 REM***************     REVISIT SUMMARY ROUTINE  ***************
2540 PRINT CHR$(147):PRINT:PRINT:REM          <CLEARS SCREEN!!!!!
2545 PRINT"ELEMENTAL SUMMARY REQUIRED (Y/N)?"
2550 GET M$:IF M$<>"Y" AND M$<>"N" THEN 2550
2555 IF M$="Y" THEN 2165
2560 PRINT"END OF PROGRAM..."
2565 GOTO 9999


2570 REM***************     PAUSE UNTIL INPUT ROUTINE  ***************
2575 PRINT:PRINT"       HIT 'RETURN' TO CONTINUE"
2580 GET A$:IF A$<>CHR$(13) THEN 2580
2585 RETURN
9999 END
```

If you wish to read analysis information from disk instead of tape then line 1155 must be altered. For PET disks line 1155 would need to be changed to:

1155 OPEN 2, 8, 3, "O:"+NA$+",S,R"

```
!-------------------------------------!
! BUILDING TYPE:FACTORY               !
!-------------------------------------!
! LOCATION      :WARWICK              !
!-------------------------------------!
! TENDER DATE   :21/02/81             !
!-------------------------------------!
! TENDER INDEX  :150                  !
!-------------------------------------!
! CONTRACT      :JCT STANDARD         !
!-------------------------------------!
! FLOOR AREA    :3000                 !
!-------------------------------------!
! NO OF STOREYS:2                     !
!-------------------------------------!

      HIT 'RETURN' TO CONTINUE
```

Screen 1

After searching for the name of the analysis file on tape it reads the file data into the main computer store and presents the front sheet information in 'boxes' on the screen.

```
ELEMENT                 EUR      COST/M2
---------------------------------------
SUBSTRUCTURE           17.50      16.80

FRAME                  39.20      39.20

UPPER FLOOR            16.40       8.88

ROOF                   29.60      28.40

PRELIMINARIES          12.32      12.32

      HIT 'RETURN' TO CONTINUE
```

Screen 2

Only appears if requested. The elemental data is presented on the screen in 'pages' if exceeding ten elements.

```
   COST PLAN BASED UPON EXAMPLE ANALYSIS

PLEASE GIVE THE FOLLOWING INFORMATION

ON THE PROJECT TO BE COST PLANNED


GROSS FLOOR AREA OF PROJECT(M2)? 2000

CURRENT INDEX VALUE           ? 175

TENDER INDEX FORECAST         ? 200

PROJECT NAME? TEST ANALYSIS

TODAYS DATE ? 02/06/82
```

Screen 3

Requests the general information required for conventional adjustment of analysis information for a cost plan.

Fig. 11.8 Major screen displays for 'Cost plan'.

Fig. 11.8 Contd.

```
COST PLAN BASED ON EXAMPLE ANALYSIS

ELEMENT NO  1         SUBSTRUCTURE

               ANALYSIS DATA
==========================================

ELEMENT UNIT RATE=              17.5

EUR UPDATED TO TODAYS DATE=     20.42

COST/M2 OF G F A =              16.8

COST/M2 @ TODAYS DATE=          19.6

==========================================
               PROJECT DATA

ELEMENT UNIT QUANTITY(PROJECT)? 2000

SPECIFICATION CHANGE(% + OR -)? 5
```

Screen 4

Presents updated element unit rate and cost per square metre and requests the measured quantity for the new job and the percentage change in the cost of the specification.

```
      SUMMARY OF ELEMENTAL COSTS

            ELEMENT NO 1

            SUBSTRUCTURE
ELEMENT UNIT QUANTITY          2000
ELEMENTAL UNIT RATE(CURRENT)   21.44
ELEMENTAL UNIT RATE(TENDER)    24.5

ELEMENTAL COST/M2(CURRENT)     21.44
ELEMENTAL COST/M2(TENDER)      24.5

TOTAL COST OF ELEMENT(TENDER) 49000

RUNN'G ELEMENTS TOT. COST/M2    24.5
RUNNING TOTAL OF ELEMENTS      49000

ARE YOU HAPPY WITH THESE FIGURES(Y/N)
```

Screen 5

Presents a summary of the costs adjusted for quantity, quality, price and location at current and tender price levels.
Allows operator to revise the element before continuing.

Fig. 11.8 Contd.

```
    ELEMENTAL COST SUMMARY

ELEMENT                 EUR        COST/M2
-------------------------------------
SUBSTRUCTURE           20.42       20.42

FRAME                  45.73       22.87

UPPER FLOOR            19.13        1.91

ROOF                   34.53       34.53

PRELIMINARIES          11.96       11.96

      HIT 'RETURN' TO CONTINUE
```

Screen 6

Summary of element target costs at either current or new project tender date levels.

```
=========================================
TEST ANALYSIS                   02/06/82

FLOOR AREA 2000        INDEX 175

        COST/M2 SUMMARY

                        POUNDS

COST/M2(CURRENT PRICES)          91.69

TOTAL COST(DITTO)         183380

COST/M2(TENDER DATE)            104.8

TOTAL COST(DITTO)         209600

=========================================
DO YOU WANT ANOTHER SUMMARY(Y/N)
```

Screen 7

Summary of total cost at current or new project tender date levels. Further screen displays allow revision of single elements and new summaries to be presented.

Developer's budget (residual valuation technique)

Introduction

This program calculates either the residual value of the land or the profit for a single land use development option. It uses the conventional residual valuation technique based on the traditional development formula:

Gross development value = net income × year's purchase

The year's purchase is calculated within the program based on the input yield and is considered in perpetuity.

Aim and objective

The aim is to demonstrate how the computer can be used to quickly evaluate a development option and then undertake a sensitivity analysis based on specified changes in the major variables. Within a period of a few minutes the user can establish the significance of the variables and gain experience of the major factors affecting that project. The program is based on the technique outlined in a number of textbooks in valuation and we are indebted to the ideas set out in a program developed by Dick Warren of The Department of Surveying, Bristol Polytechnic, Bristol, UK.

Scope of program

Certain simplifications have been made for the sake of clarity including:

(i) Printing of hardcopy has not been included.
(ii) Reduction in the possible number of items under cost of development.
(iii) Input checks have been kept to a minimum.
(iv) A valuation summary is not included.
(v) The year's purchase table is in perpetuity.

Machine requirements
(i) Central processing unit/keyboard
(ii) Screen
(iii) Cassette tape or disk for recording the program.

Data requirements

The operator will need to have before him

(i) A knowledge of the expected rental, floor area, expected yield etc.
(ii) Details of the initial and annual costs relating to the development. In particular he will need to know the construction costs including an allowance for inflation.
(iii) Details such as floor area of the project under consideration.

Possible alterations and development

The program represents a simple introduction to the residual valuation technique. There are a number of ways in which it can be developed including:

(i) Hardcopy of intermediate results, final results, and valuation details.
(ii) Extension of routines to cope with a mixed development as opposed to a single use development.
(iii) Automatic sensitivity analysis routines which test the effect of changing one variable at a time in incremental steps.
(iv) Recording of results on to disk or tape.
(v) The opportunity to print out full details of any sensitivity analysis.

Note: The program is written for a Commodore PET. The few instructions which may differ for other machines are included in table 11.1.

There follow the flow chart (fig. 11.9), the program listing, and the major screen displays (fig. 11.10) for 'Developer's budget'.

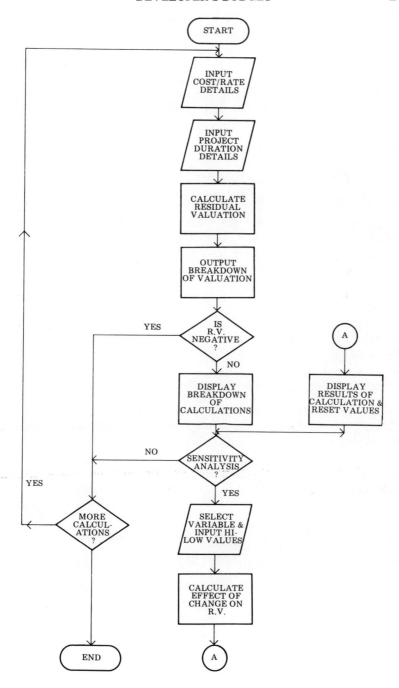

Fig. 11.9 Flow chart for 'Developer's budget'.

Program listing
Developer's budget

```
1000 REM*************** DEVELOPER'S BUDGET PROGRAM ***************

1005 REM********(C) COPYRIGHT BRANDON,MOORE & MAIN. 20-05-82********
1010 PRINT CHR$(147);:REM                                 <CLEARS SCREEN!!!!
1015 PRINT"     DEVELOPMENT  FEASIBILITY"
1020 PRINT"     ======================="
1025 PRINT"          ANALYSIS"
1030 PRINT"          =======":PRINT:PRINT
1035 PRINT"     A RESIDUAL VALUATION PROGRAM":PRINT
1040 PRINT"        WITH OPTIONS FOR":PRINT
1045 PRINT"     SENSITIVITY ANALYSIS":PRINT
1050 GOSUB2305:REM                                        <HALT UNTIL INPUT
1055 CLR
1060 LN$="======================================="
1065 PRINT CHR$(147);:REM                                 <CLEARS SCREEN!!!!
1070 PRINT"  KEY IN EXPECTED VALUES FOR PROJECT":PRINT:PRINT
1075 INPUT"     RENT IN POUNDS PER  SQ.M";I1:W=I1:PRINT:PRINT
1080 INPUT"     GROSS FLOOR AREA IN SQ.M";A1:B=A1:PRINT:PRINT
1085 INPUT"     % OF GROSS AREA LETTABLE";K1:K=K1/100:A2=A1*K:PRINT:PRINT
1090 INPUT"     ANNUAL OUTGOINGS        ";CA:PRINT:PRINT
1095 INPUT"     MARKET YIELD            ";Y1
1100 IF Y1=0 THEN PRINT"ERROR":GOTO 1095
1105 Y=Y1/100:YP=1/Y:PRINT:PRINT


1110 REM***************     GDV CALC        ***************
1115 M=B*K*W:IN=M-CA:V=IN*YP
1120 PRINT"   IS SITE VALUE KNOWN(Y/N)"
1125 GET A$:IF A$<>"Y" AND A$<>"N" THEN 1125
1130 PRINT CHR$(147);:REM                                 <CLEARS SCREEN!!!!
1135 IF A$="Y" THEN PRINT"CALCULATION OF RESIDUAL PROFIT":Z=1
1140 IF A$="N" THEN PRINT"CALCULATION OF RESIDUAL SITE VALUE"
1145 PRINT:PRINT"  COST DATA REQUIRED:-":PRINT


1150 REM************** ROUTINE INPUT PROJ.DETAILS  ***************
1155 IF Z=0 THEN INPUT"DEVELOPER PROFIT AS % OF COSTS";DP:O=DP:PRINT:GOTO 1165
1160 INPUT"COST OF SITE, PLUS FEES ETC.  ";DP:PRINT:O=0:Q=DP
1165 PRINT"CONSTRUCTION COST PER SQ.M."
1170 INPUT"ALLOWING FOR INFLATION      ";CC:C=CC:PRINT
1175 INPUT"CLEARANCE & SITE WORKS      ";CS:D=CS:PRINT
1180 INPUT"% DESIGN FEES ON CONSTRUCTION ";E1:PRINT
1185 INPUT"% CONTINGENCIES    DITTO    ";CG:G=CG:PRINT
1190 INPUT"DISPOSAL EXPENSES(% NETT INC.)";FE:PRINT
1195 INPUT"BORROWING RATE OF INTEREST  ";S2:S1=S2/100:S=S1+1:R=S:PRINT
1200 INPUT"OTHER COSTS                 ";P1
1205 PRINT CHR$(147);:REM                                 <CLEARS SCREEN!!!!
1210 PRINT"   ESTIMATED DURATION OF PROJECT":PRINT:PRINT
1215 PRINT"TIME FROM SITE ACQUISITION      ":PRINT
1220 INPUT"TO START OF CONSTRUCTION.(MONTHS)";SM:SY=SM/12:PRINT:PRINT:PRINT
1225 INPUT"CONSTRUCTION PERIOD.    (MONTHS)";TM:TY=TM/12:T1=TY/2
1230 PRINT:PRINT:PRINT
```

Program listing
Developer's budget Contd.

```
1235 INPUT"LETTING VOID.           (MONTHS)";LM:'LY=LM/12
1240 GOSUB2000:REM                                          <RESIDUAL CALCULTN


1245 REM***************   INITIAL CALC. PRINTOUT   ***************
1250 PRINT CHR$(147);:REM                                   <CLEARS SCREEN!!!!
1255 PRINT"     RESULTS OF THIS CALCULATION"
1260 PRINT"     ==========================="
1265 PRINT"                              POUNDS"
1270 PRINT"                              ------"
1275 V=INT(((V+0.5)*10)/10)
1280 PRINT" GROSS DEVELOPMENT VALUE";
1285 ZZ=V:GOSUB 2330:PRINT
1290 PRINT" TOTAL COSTS          ";
1295 T=INT(((T+0.5)*10)/10)
1300 ZZ=T:GOSUB 2330:PRINT
1305 PRINT"                        =============="
1310 RV=INT(((RV+0.5)*10)/10)
1315 IF Z=0 THEN 1345
1320 ZZ=RV:IF RV>0 THEN 1330
1325 PRINT" RESIDUAL LOSS        ";:GOSUB 2330:PRINT:GOTO 1335
1330 PRINT" RESIDUAL PROFIT      ";:GOSUB 2330:PRINT
1335 PRINT"                        =============="
1340 PRINT:GOSUB 2305:GOTO 1510
1345 PRINT" RESIDUAL VALUE       ";
1350 IF RV<0 THEN PRINT:PRINT:PRINT"   RESIDUAL SITE VALUE IS NEGATIVE."
1355 IF RV<0 THEN GOSUB 2305:GOTO 1970
1360 ZZ=RV:GOSUB 2330:PRINT
1365 PRINT"                        =============="
1370 SC=V1/1.04
1375 SC=INT(((SC+0.5)*10)/10):PRINT
1380 PRINT LN$
1385 FOR W1=1 TO 7:PRINT"!                              !":NEXT W1
1390 PRINT LN$
1395 PRINT CHR$(19);:REM                                    <HOMES SCREEN!!!!
1400 FOR QQ=1 TO 11:PRINT:NEXT QQ
1405 PRINT"!  MAXIMUM SITE COSTS    ";
1410 ZZ=SC:GOSUB 2330:PRINT
1415 PRINT"!  ------------------"
1420 PRINT"!  PLUS":PRINT
1425 PRINT"!  4% LEGAL COSTS       ";
1430 SZ=INT((((SC*.04)+0.5)*10)/10)
1435 ZZ=SZ:GOSUB 2330:PRINT:REM                             <FORMAT LEGAL COSTS
1440 PRINT"!  AND INTEREST "
1445 PRINT"! @ ";
1450 ZZ=S2:SP=2:GOSUB 2335:REM                              <FORMAT % RATE
1455 PRINT"% FOR";
1460 ZZ=SM+TM+LM:SP=1:GOSUB 2335:REM                        <FORMAT TIME PERIOD
1465 PRINT"MONTHS";
1470 RZ=INT((((RV-V1)+0.5)*10)/10)
1475 ZZ=RZ:GOSUB 2330:PRINT:REM                             <FORMAT INTEREST
1480 PRINT LN$
1485 PRINT:PRINT" DEV.PROF @";
1490 ZZ=DP:SP=1:GOSUB 2335
1495 PRINT"% OF COST";
```

Program listing
Developer's budget Contd.

```
1500 ZZ=INT(((N+0.5)*10)/10):GOSUB 2330:PRINT:REM          <FORMAT PROFIT
1505 GOSUB 2305:REM                                        <HALT UNTIL INPUT
1510 PRINT CHR$(147);:REM                                  <CLEARS SCREEN!!!!
1515 PRINT:PRINT:PRINT:PRINT:PRINT:PRINT:PRINT
1520 PRINT"CHANGE INPUT VALUES(Y/N)"
1525 GET A$:IF A$<>"Y" AND A$<>"N" THEN1525
1530 IF A$="Y" THEN 1055
1535 PRINT CHR$(147);:REM                                  <CLEARS SCREEN!!!!
1540 PRINT:PRINT:PRINT:PRINT:PRINT:PRINT:PRINT
1545 PRINT"TYPE 'P' FOR A COPY OF THE VALUATION"
1550 PRINT:PRINT"('RETURN' TO CONTINUE)"
1555 GET A$:IF A$<>"P" AND A$<>CHR$(13) THEN 1555
1560 IF A$="P" THEN GOSUB 2035
1565 GOTO 1935


1570 REM*************** SENSITIVITY ANALYSIS ROUTINE ***************
1575 PRINT CHR$(147);:REM                                  <CLEARS SCREEN!!!!
1580 PRINT"        SENSITIVITY ANALYSIS":PRINT
1585 PRINT"WHILE  EXPECTED VALUES REMAIN CONSTANT"
1590 PRINT"ONE VARIABLE IS CHANGED."
1595 PRINT"THE EFFECT ON RESID. VALUE IS ANALYSED.":PRINT
1600 PRINT"     WHICH VARIABLE TO BE CHANGED"
1605 PRINT"     ---------------------------":PRINT
1610 PRINT"     RENT.......................R ":PRINT
1615 PRINT"     CONST. COST PER SQ.M......C ":PRINT
1620 PRINT"     BORROWING RATE............I ":PRINT
1625 IF Z=1 THEN1635
1630 PRINT"     PROFIT % OF COSTS.........P ":PRINT
1635 PRINT"     MARKET YIELD..............Y ":PRINT
1640 INPUT"     ENTER RELEVANT LETTER    ";C1$
1645 IF C1$="R" THEN C$="RENT.    ":D$="POUNDS/SQ.M":XX=W:GOTO 1675
1650 IF C1$="C" THEN C$="CON.COST":D$="POUNDS/SQ.M":XX=CC:GOTO 1675
1655 IF C1$="I" THEN C$="INT.RATE":D$="%":XX=S2:GOTO 1675
1660 IF C1$="P" AND Z=0 THEN C$="PROFIT  ":D$="% OF COSTS":XX=DP:GOTO 1675
1665 IF C1$="Y" THEN C$="YIELD.  ":D$="%":XX=Y1:GOTO 1675
1670 INPUT"        ERROR! TRY AGAIN    ";C1$:GOTO 1645
1675 PRINT CHR$(147);:REM                                  <CLEARS SCREEN!!!!
1680 PRINT:PRINT:PRINT
1685 PRINT"    ANALYSIS OF VARIABLE- ";C$
1690 PRINT"    ================================="
1695 PRINT:PRINT:PRINT
1700 PRINT" EXPECTED ";C$;" IS";XX;D$:PRINT:PRINT
1705 PRINT" ENTER TWO DIFFERENT VALUES FOR ";C$:PRINT:PRINT
1710 INPUT"     HIGH VALUE";HI:PRINT:PRINT
1715 INPUT"     LOW VALUE ";LO


1720 REM*************** ROUTINE FOR ANALYSIS RESULTS ***************
1725 PRINT CHR$(147);:REM                                  <CLEARS SCREEN!!!!
1730 PRINT:PRINT:PRINT"   RESULTS OF SENSITIVITY ANALYSIS"
1735 PRINT"   ============================="
1740 PRINT:PRINT"      CHANGED VARIABLE ";C$
1745 PRINT:PRINT:PRINTC$;"  G.D.V.     COSTS     RESIDUAL"
1750 Z$="PROFIT  ":IF Z=0 THEN Z$="SITE VAL"
```

Program listing
Developer's budget Contd.

```
1755 FOR JJ=1 TO 5
1760 PRINT CHR$(19);:REM                                      <HOMES SCREEN!!!!
1765 FOR QQ=1 TO 9:PRINT:NEXT QQ:REM                          <CURSOR DOWN
1770 PRINT"                                      ":REM        <CAUSE FLASH
1775 PRINT CHR$(19);:REM                                      <HOMES SCREEN!!!!
1780 FOR QQ=1 TO 9:PRINT:NEXT QQ:REM                          <CURSOR DOWN
1785 FOR QQ=1 TO 200:NEXT QQ:PRINT"                 ";Z$
1790 FOR QQ=1 TO 200:NEXT QQ:NEXT JJ


1795 REM*************** CALCULATION OF SENSITIVITY   ***************
1800 IF C1$="C" THEN M=B*K*W:IN=M-CA:V=IN*YP:PRINT:REM        <GDV CALCULATION
1805 IF C1$="R" THEN W=HI
1810 IF C1$="C" THEN C=HI
1815 IF C1$="I" THEN S5=HI/100:R=S5+1
1820 IF C1$="P" THEN O=HI
1825 IF C1$="Y" THEN Y3=HI/100:Y2=1/Y3:YP=Y2
1830 M=B*K*W:IN=M-CA:V=IN*YP:REM                              <GDV CALCULATION
1835 GOSUB 2000:REM                                           <RESIDUAL CALCULATN
1840 SP=2:ZZ=HI:PRINT:II=1:GOSUB 2345:II=0
1845 SP=8:ZZ=V:GOSUB 2345
1850 SP=8:ZZ=T:GOSUB 2345
1855 SP=8:ZZ=RV:GOSUB 2345:PRINT
1860 IF C1$="R" THEN W=LO
1865 IF C1$="C" THEN C=LO
1870 IF C1$="I" THEN S8=LO/100:R=S8+1
1875 IF C1$="P" THEN O=LO
1880 IF C1$="Y" THEN Y3=LO/100:Y2=1/Y3:YP=Y2
1885 M=B*K*W:IN=M-CA:V=IN*YP:REM                              <GDV CALCULATION
1890 GOSUB 2000:REM                                           <RESIDUAL CALCULATN
1895 SP=2:ZZ=LO:PRINT:II=1:GOSUB 2345:II=0
1900 SP=8:ZZ=V:GOSUB 2345
1905 SP=8:ZZ=T:GOSUB 2345
1910 SP=8:ZZ=RV:GOSUB 2345:PRINT
1915 GOSUB 2305:REM                                           <HALT UNTIL INPUT


1920 REM***************    RESET EXPECTED VALUES    ***************
1925 B=A1:C=CC:D=CS:G=CG:R=S:O=DP:W=I1:YP=1/Y


1930 REM*************** Y/N ROUTINE FOR MORE ANALYSIS***************
1935 PRINT CHR$(147);:REM                                     <CLEARS SCREEN!!!!
1940 PRINT:PRINT:PRINT:PRINT:PRINT:PRINT:PRINT
1945 PRINT" SENSITIVITY ANALYSIS (Y/N)"
1950 GET A$:IF A$<>"Y" AND A$<>"N" THEN 1950
1955 IF A$="Y" THEN 1575
1960 GOTO 1970


1965 REM*************** Y/N ROUTINE FOR CALCULATION ***************
1970 PRINT CHR$(147);:REM                                     <CLEARS SCREEN!!!!
1975 PRINT:PRINT:PRINT:PRINT:PRINT:PRINT:PRINT
1980 PRINT"ANOTHER CALCULATION (Y/N)"
1985 GET A$:IF A$<>"Y" AND A$<>"N" THEN 1985
```

Program listing
Developer's budget Contd.

```
1990 IF A$="N" THEN 2400
1995 GOTO 1055


2000 REM****************    RESIDUAL CALCULATION.    ****************
2005 A=(B*C)+D:E=A*E1/100:F=A*G/100:H=A+F+E+P1
2010 J=H*R↑(T1+LY)-H:L=M*FE/100:IF Z=1 THEN 2020
2015 N=A*O/100:T=H+J+L+N:RV=V-T:V1=RV-(RV*(R↑(SY+TY+LY))-RV):GOTO 2025
2020 P=Q/(2-(R↑(SY+TY+LY)))-Q:T=H+J+Q+L+P:RV=V-T
2025 RETURN


2030 REM***************    DISPLAY RESULTS    ***************
2035 PRINT CHR$(147);:REM                                  <CLEARS SCREEN!!!!
2040 PRINT"RESIDUAL VALUATION":PRINT
2045 IF Z=0 THEN 2055
2050 PRINT"SITE COST                    =";:ZZ=DP:GOSUB 2330:PRINT:PRINT
2055 PRINT"GROSS INCOME                 =";:ZZ=M:GOSUB 2330:PRINT
2060 PRINT"OUTGOINGS                    =";:ZZ=CA:GOSUB 2330:PRINT
2065 PRINT"                 ------------"
2070 PRINT"NET INCOME                   =";:ZZ=IN:GOSUB 2330:PRINT
2075 PRINT"                 ------------"
2080 GOSUB 2305:REM                                       <HALT UNTIL INPUT
2085 PRINT CHR$(147);:REM                                  <CLEARS SCREEN!!!!
2090 PRINT"Y.P. (PERP) @ ";:ZZ=Y1:SP=2:GOSUB 2335:PRINT"%=";
2095 ZZ=YP:GOSUB 2335:PRINT
2100 PRINT"GROSS DEVELOPMENT VALUE  ";:ZZ=V:GOSUB 2330:PRINT:PRINT
2105 PRINT"CONSTRUCTION COST        ";:ZZ=B*C:GOSUB 2330:PRINT
2110 PRINT"SITE WORKS               ";:ZZ=D:GOSUB 2330:PRINT
2115 PRINT LN$
2120 PRINT"TOTAL CONSTRUCTION COSTS ";:ZZ=D+B*C:GOSUB 2330:PRINT
2125 PRINT LN$
2130 GOSUB 2305:REM                                       <HALT UNTIL INPUT
2135 PRINT CHR$(147);:REM                                  <CLEARS SCREEN!!!!
2140 PRINT"TOTAL CONSTRUCTION COSTS ";:ZZ=D+B*C:GOSUB 2330:PRINT
2145 PRINT"FEES @ ";:ZZ=E1:SP=2:GOSUB 2335:PRINT"%";
2150 ZZ=E:GOSUB 2330:PRINT
2155 PRINT"CONTINGENCIES @ ";:ZZ=G:SP=2:GOSUB 2335:PRINT"%";
2160 ZZ=F:GOSUB 2330:PRINT
2165 T2=INT((12*(T1+LY))+0.5)
2170 PRINT"FINANCE";:ZZ=S2:SP=2:GOSUB 2335:PRINT"% FOR";
2175 ZZ=T2:SP=2:GOSUB 2335:PRINT"MTH";
2180 ZZ=J:GOSUB 2330:PRINT
2185 PRINT"LETTING.EXPENSES         ";:ZZ=L:GOSUB 2330:PRINT
2190 PRINT"DEVELOPERS PROFIT  ";:ZZ=O:SP=2:GOSUB 2335:PRINT"%";
2195 ZZ=N:GOSUB 2330:PRINT
2200 PRINT"-------------------------------------"
2205 PRINT"TOTAL COSTS              ";:ZZ=T:GOSUB 2330:PRINT
2210 IF Z=1 THEN 2275
2215 PRINT LN$
2220 PRINT"RESIDUAL VALUE           ";:ZZ=RV:GOSUB 2330:PRINT
2225 PRINT LN$
2230 PRINT:PRINT"APPORTIONED AS FOLLOWS":PRINT
2235 PRINT"MAXIMUM SITE COST        ";:ZZ=SC:GOSUB 2330:PRINT
2240 PRINT"INTEREST";:ZZ=S2:SP=2:GOSUB 2335:PRINT"OVER";
```

Program listing
Developer's budget Contd.

```
2245 ZZ=SM+TM+LM:SP=2:GOSUB 2335:PRINT"MTH";
2250 ZZ=RV-V1:GOSUB 2330:PRINT
2255 PRINT"LEGAL COSTS OF ACQ.      ";:ZZ=SC*0.04:GOSUB 2330:PRINT
2260 PRINT LN$
2265 PRINT"RESIDUAL VALUE           ";:ZZ=RV:GOSUB 2330:PRINT
2270 GOTO 2285
2275 PRINT LN$
2280 PRINT"RESIDUAL PROFIT          ";:ZZ=RV:GOSUB 2330:PRINT
2285 PRINT LN$
2290 GOSUB 2305:REM                                <HALT UNTIL INPUT
2295 RETURN


2300 REM*************** SUBROUTINE PAUSE/CONTINUE ***************
2305 FOR QQ=1 TO 1000:NEXT QQ:PRINT
2310 PRINT"     PRESS 'RETURN' TO CONTINUE"
2315 GET ZZ$:IF ZZ$<>CHR$(13) THEN 2315
2320 RETURN


2325 REM************** SUBROUTINES TO FORMAT NUMBER ***************
2330 PRINT TAB(36-LEN(STR$(INT(ZZ))));INT(ZZ);:RETURN:REM <TAB METHOD
2335 PRINT INT(0.5+100*ZZ)/100;:RETURN:REM               <NO TO 2 DPS


2340 REM*************** ROUTINE USING 'SPACE' METHOD ***************
2345 ZY=ZZ:REM                                    <COPY NUMBER
2350 IF ABS(ZZ)>0 AND ABS(ZZ)<1 THEN SP=11:GOTO 2365:REM <DEALS WITH (0<NO<1)
2355 IF ABS(ZY/10)-1>=0 THEN ZY=ZY/10:SP=SP-1:GOTO 2355
2360 IF SP=0 THEN 2385
2365 FOR QQ=1 TO SP:REM                           <PRINT SPACES
2370 PRINT" ";
2375 NEXT QQ
2380 IF II=1 THEN PRINT INT(.5+100*ZZ)/100;:GOTO 2390:REM <NO TO 2 DPS
2385 PRINT INT(ZZ);:REM                           <PRINT NUMBER
2390 RETURN


2395 REM***************     END OF PROGRAM     ***************
2400 PRINT CHR$(147);:REM                         <CLEARS SCREEN!!!!
2405 PRINT"END OF PROGRAM"
9999 END
```

```
KEY IN EXPECTED VALUES FOR PROJECT

  RENT IN POUNDS PER  SQ.M? 40

  GROSS FLOOR AREA IN SQ.M? 2000

  % OF GROSS AREA LETTABLE? 90

  ANNUAL OUTGOINGS        ? 2500

  MARKET YIELD            ? 5.5

  IS SITE VALUE KNOWN(Y/N)
```

Screen 1

Input the revenue details of new development.

Asks whether site value is known. If the answer is yes the residual value is the site. If not the program calculates the profit.

```
CALCULATION OF RESIDUAL SITE VALUE

  COST DATA REQUIRED:-

DEVELOPER PROFIT AS % OF COSTS? 10

CONSTRUCTION COST PER SQ.M.
ALLOWING FOR INFLATION      ? 380

CLEARANCE & SITE WORKS       ? 1500

% DESIGN FEES ON CONSTRUCTION ? 6

% CONTINGENCIES    DITTO     ? 5

DISPOSAL EXPENSES(% NETT INC.)? 5

BORROWING RATE OF INTEREST   ? 10

OTHER COSTS                  ? 0
```

Screen 2

Input details of the development costs.

```
   ESTIMATED DURATION OF PROJECT

TIME FROM SITE ACQUISITION

TO START OF CONSTRUCTION.(MONTHS)? 6

CONSTRUCTION PERIOD.    (MONTHS)? 18

LETTING VOID.          (MONTHS)? 0
```

Screen 3

Input details of design and construction times used for calculating finance.

Fig. 11.10 Major screen displays for 'Developer's budget'.

Fig. 11.10 Contd.

```
     RESULTS OF THIS CALCULATION
     ============================
                            POUNDS
                            ------
 GROSS DEVELOPMENT VALUE    1263636
 TOTAL COSTS                 987649
                        ==============
 RESIDUAL VALUE              275988
                        ==============


 ====================================
 !   MAXIMUM SITE COSTS      209644  !
 !   ------------------              !
 !   PLUS                            !
 !                                   !
 !   4% LEGAL COSTS            8386  !
 !   AND INTEREST                    !
 !   @ 10 % FOR 24 MONTHS     57958  !
 ====================================

  DEV.PROF @ 10 % OF COST     76150

      PRESS 'RETURN' TO CONTINUE
```

Screen 4

Output of residual value of land. If residual value is negative it asks whether a new calculation is required.

Fig. 11.10 Contd.

```
RESIDUAL VALUATION

GROSS INCOME              =       72000
OUTGOINGS                 =        2500
                              ------------
NET INCOME                =       69500
                              ------------

        PRESS 'RETURN' TO CONTINUE
```

Screens 5, 6 and 7

These are 'paged' screen displays which are merely summaries of the input information and results.

```
Y.P. (PERP)  @ 5.5 %= 18.18
GROSS DEVELOPMENT VALUE        1263636

CONSTRUCTION COST              760000
SITE WORKS                       1500
=======================================
TOTAL CONSTRUCTION COSTS       761500
=======================================

        PRESS 'RETURN' TO CONTINUE
```

```
TOTAL CONSTRUCTION COSTS       761500
FEES  @ 6 %                     45690
CONTINGENCIES @ 5 %             38075
FINANCE 10 % FOR 9 MTH          62633
LETTING EXPENSES                 3600
DEVELOPERS PROFIT  @ 10 %       76150
--------------------------------------
TOTAL COSTS                    987649
=======================================
RESIDUAL VALUE                 275988
=======================================

APPORTIONED AS FOLLOWS

MAXIMUM SITE COST              209644
INTEREST 10 OVER 24 MTH         57957
LEGAL COSTS OF ACQ.              8385
=======================================
RESIDUAL VALUE                 275988
=======================================

        PRESS 'RETURN' TO CONTINUE
```

Fig. 11.10 Contd.

```
        SENSITIVITY ANALYSIS

WHILE  EXPECTED VALUES REMAIN CONSTANT
ONE VARIABLE IS CHANGED.
THE EFFECT ON RESID. VALUE IS ANALYSED.

    WHICH VARIABLE TO BE CHANGED
    ----------------------------

    RENT.....................R

    CONST. COST PER SQ.M......C

    BORROWING RATE............I

    PROFIT % OF COSTS.........P

    MARKET YIELD..............Y

    ENTER RELEVANT LETTER   ? R
```

Screen 8

Provides the opportunity to undertake a simple sensitivity analysis on the major variables.

```
    ANALYSIS OF VARIABLE- RENT.
    ===================================



EXPECTED RENT.    IS 40 POUNDS/SQ.M


ENTER TWO DIFFERENT VALUES FOR RENT.

    HIGH VALUE? 45

    LOW VALUE ? 35
```

Screen 9

Outputs the current value of the variable to be tested and requests new high and low values for sensitivity.

```
    RESULTS OF SENSITIVITY ANALYSIS
    ===============================

        CHANGED VARIABLE RENT.

RENT.    G.D.V.      COSTS     RESIDUAL
                               SITE VAL

 45     1427272     988098      439174

 35     1100000     987198      112801

    PRESS 'RETURN' TO CONTINUE
```

Screen 10

Provides the new residual values for the changes in the single variable.

By this method the significance of each variable can be tested.

Heat loss

Introduction

This program contains the calculations necessary to find the total steady state heat loss from a room or from a building. It can also be used to calculate U-values for various types of construction.

Aim and objective

The aim is to provide a rapid method of calculating heat losses so that the effect of variations in methods of construction, window areas etc. can quickly be assessed.

Scope of program

The program is based on *Heating* by C. R. Bassett and D. C. Pritchard (Longman). The six tables contained in chapter 3 of that book are included in full as data in lines 6005—6045.

A list of materials with their k-values is included as data in lines 6065—6090 and provision is made for adding to this data during the running of the program. If the option at the end of the program to calculate further U-values is taken, any additions which have been made will be displayed on the screen on subsequent runs. If the program is rerun from the beginning this information will be lost.

The array $R(16, 10)$ is used to store the R-values as they are calculated. The room or building is assumed to have four walls and each wall can contain up to four different types of construction, e.g. wall, windows, doors etc. The corresponding areas of each are contained in the array $A(4, 4)$. The R-values for wall number 1 will be stored in the first four rows of the array $R(\ )$, those for wall number 2 in rows 5—8 etc. The ten values in the columns are used to store a maximum of nine R-values starting from the outside. The tenth value is a count of the number of R-values used in each column. The U-values for the walls are contained in the array $U(4, 4)$ and those for the floor and ceiling in $U1(2)$.

Machine requirements
  (i) Central processing unit/keyboard

(ii) Screen

(iii) Cassette tape deck and blank tape or disk drive and disk for storing the program.

Data requirements

The operator will need to have before him for entry via the keyboard, the materials and thicknesses used in the construction and the particulars of any cavities. If it is an external wall its orientation and the conditions of exposure must also be known. If any of the materials used are not contained in the data bank it is also necessary to have the k-values of these items. If the option to calculate the steady state heat loss is taken, the dimensions of the room and those of the various types of construction together with internal and external temperatures and the rate of air change must also be known.

Possible alterations and development

(i) Hard copy of results — to include the data supplied, e.g. materials and thicknesses etc. and the calculated U-values. If the option for steady state heat loss is chosen, the information leading to this and the final result can also be included.

(ii) Use a data file to contain all the information now in the data lines. Particulars of any new materials and their k-values added during a run of the program can then be stored in this file for future use.

(iii) Addition of a sort routine to ensure that the list of materials remains in alphabetical order when new materials are added.

(iv) Extend the program to include the calculation of the temperature gradient and hence pinpoint any possible intersticial condensation problems. The R-values have been held in the array R( ) for this purpose.

(v) Change the very simple input routines which have been used in this program to present a better appearance on the screen.

(vi) If the steady state heat loss is being calculated, once the orientation of one outside wall has been given the machine can be instructed how to obtain the orientation of any of the remaining three walls which may be external.

There follow the flow chart (fig. 11.11), the program listing, and the major screen displays (fig. 11.12) for 'Heat loss'.

284

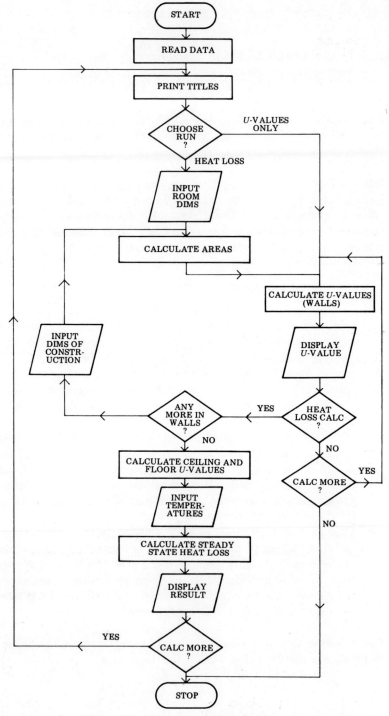

Fig. 11.11 Flow chart for 'heat loss'.

Program listing
Heat loss

```
1000 REM************HEATLOSS STARTER PACK 18/03/82***********


1001 REM*********(C)COPYRIGHT BRANDON,MOORE & MAIN***********
1005 DIM A(4,4),D$(30),T(4),T1(5,3),T2(4),T3(9),T4(6,2):REM<DIM VARIABLES
1010 DIM T5(6,3),T6(4),K(30),R(16,10),U(4,4),U1(2)


1015 REM************  READ VALUES INTO TABLE 1 ************
1020 FOR I=1 TO 5
1025 FOR J=1 TO 3
1030 READ T1(I,J)
1035 NEXT J,I


1040 REM*************READ VALUES INTO TABLES 2 & 6***********
1045 FOR I=1 TO 4
1050 READ T2(I)
1055 READ T6(I)
1060 NEXT I


1065 REM************  READ VALUES INTO TABLE 3 ************
1070 FOR I=1 TO 9
1075 READ T3(I)
1080 NEXT I


1085 REM************  READ VALUES INTO TABLE 4 ************
1090 FOR I=1 TO 6
1095 FOR J=1 TO 2
1100 READ T4(I,J)
1105 NEXT J,I


1110 REM************  READ VALUES INTO TABLE 5 ************
1115 FOR I=1 TO 6
1120 FOR J=1 TO 3
1125 READ T5(I,J)
1130 NEXT J,I


1135 REM************SET'N' TO NO. OF VALUES OF K************
1140 N=18
1145 TL=500:REM                                 <TIME LAG


1150 REM************ READ K VALUES & MATERIALS  ***********
1155 FOR I=1 TO N
1160 READ D$(I),K(I)
1165 NEXT I
1170 PRINT CHR$(147);:REM                       <CLEARS SCREEN!!!!
1175 PRINT"HEAT LOSS CALCULATIONS"
```

Program listing
Heat loss Contd.

```
1180 PRINT"======================="
1185 PRINT"THIS PROGRAM IS FOR THE CALCULATION OF"
1190 PRINT"U VALUES FOR VARIOUS CONSTRUCTIONS.":PRINT
1195 PRINT"THE PROGRAM WILL CONTINUE IF DESIRED"
1200 PRINT"TO CALCULATE THE STEADY STATE HEAT LOSS"
1205 PRINT"FROM A RECTANGULAR ROOM.":PRINT
1210 PRINT"ARE CALCULATIONS REQUIRED OF:-":PRINT
1215 PRINT"1) U-VALUES ONLY.":PRINT
1220 PRINT"2) U-VALUES & HEATLOSS.":PRINT
1225 PRINT"SELECT OPTION REQUIRED."
1230 GET A$:IF A$<>"1" AND A$<>"2" THEN 1230
1235 Q=VAL(A$):PRINT:PRINT A$;" SELECTED.":FOR X=1 TO TL:NEXT X
1240 PRINT CHR$(147);:REM                          <CLEARS SCREEN!!!!
1245 IF Q=1 THEN 1335
1250 PRINT"ROOM DIMENSIONS (M)":PRINT
1255 INPUT"LENGTH";L
1260 INPUT"WIDTH ";W
1265 INPUT"HEIGHT";H
1270 A(1,1)=L*H
1275 A(3,1)=L*H
1280 A(2,1)=W*H
1285 A(4,1)=W*H
1290 V=L*W*H
1295 A1=L*W


1300 REM************* CALCULATE U VALUES        ************
1305 PRINT:PRINT"WALLS 1 & 3 ARE CONSIDERED TO BE"
1310 PRINT"IN THE LENGTH OF THE ROOM."
1315 PRINT"WALLS 2 & 4 ARE CONSIDERED TO BE"
1320 PRINT"IN THE WIDTH OF THE ROOM.":PRINT
1325 PRINT"TOUCH 'RETURN' TO CONTINUE."
1330 GET Z$:IF Z$<>CHR$(13) THEN 1330
1335 N1=1
1340 PRINT CHR$(147);:REM                          <CLEARS SCREEN!!!!
1345 PRINT"DETAILS OF WALL CONSTRUCTION":PRINT
1350 IF Q=1 THEN 1360
1355 PRINT"WALL NUMBER";N1
1360 PRINT"IS THIS AN EXTERNAL WALL? (Y/N)":PRINT
1365 GET W$:IF W$<>"Y" AND W$<>"N" THEN 1365
1370 IF W$="N" THEN 1445


1375 REM************* ADAPTION FOR EXTERNAL WALL ************
1380 PRINT"WALL FACING"
1385 PRINT"1) S"
1390 PRINT"2) W, S.W , S.E"
1395 PRINT"3) N.W"
1400 PRINT"4) N, N.E , E"
1405 GET Z$:F=VAL(Z$):IF F=0 OR F>4 THEN 1405
1410 PRINT:PRINT Z$;" SELECTED":PRINT
1415 PRINT"CONDITIONS OF EXPOSURE"
1420 PRINT"1) SHELTERED."
1425 PRINT"2) NORMAL."
1430 PRINT"3) SEVERE."
```

Program listing
Heat loss Contd.

```
1435 GET Z$:E=VAL(Z$):IF E=0 OR E>3 THEN 1435
1440 PRINT:PRINT Z$;" SELECTED":FOR X=1 TO TL:NEXT X
1445 N3=1
1450 N2=1
1455 I=(N1-1)*4+N3
1460 D=0:R(I,10)=1:R(I,1)=T1(F,E):IF W$="N" THEN R(I,1)=T2(1)
1465 GOSUB 1495:REM                              <CALC U-VALUE ROUTINE
1470 IF Q=2 THEN 1485
1475 IF Q$="Y" THEN 1240
1480 IF Q$="N" THEN 9999
1485 IF Q$="N" THEN 2110
1490 GOTO 1450
1495 PRINT CHR$(147);:REM                        <CLEARS SCREEN!!!!

1500 REM************* DISPLAY MATERIALS AVAILABLE************
1505 PRINT"MATERIALS ARE:-"
1510 FOR J=1 TO 30 STEP 2
1515 PRINT STR$(J);")";D$(J);TAB(20);STR$(J+1);")";D$(J+1)
1520 IF J<N THEN 1530
1525 J=30
1530 NEXT J
1535 PRINT
1540 IF N2<>1 THEN 1555
1545 PRINT"MATERIALS TO BE ENTERED STARTING"
1550 PRINT"FROM OUTSIDE ROOM"
1555 PRINT" MATERIAL NUMBER";N2
1560 PRINT"(CAVITY COUNTS AS A MATERIAL)"
1565 N2=N2+1
1570 PRINT"IF MATERIAL IS IN LIST ENTER NUMBER"
1575 PRINT"IF NEW MATERIAL IS TO BE ADDED ENTER 50"
1580 INPUT"MATERIAL CODE ";X:REM                 <SELECT MATERIAL
1585 IF X=50 THEN 1900
1590 IF X=0 OR X>N THEN PRINT"ERROR":GOTO 1580
1595 INPUT"THICKNESS (MM)";T
1600 IF D=1 THEN 1620
1605 R(I,N2)=T/K(X)/1E3
1610 R(I,10)=R(I,10)+1
1615 GOTO 1625
1620 R1=R1+T/K(X)/1E3
1625 PRINT CHR$(147);:REM                         <CLEARS SCREEN!!!!
1630 PRINT"ENTER '1' FOR NEXT MATERIAL."
1635 PRINT"ENTER '2' FOR CAVITY."
1640 PRINT"ENTER '3' IF COMPLETE."
1645 GET Z$:NO=VAL(Z$):IF NO=0 OR NO>3 THEN 1645
1650 ON NO GOTO 1495,1660,1940

1655 REM*************     ROUTINE FOR CAVITIES     ************
1660 N2=N2+1:PRINT
1665 PRINT"1) VENTILATED."
1670 PRINT"2) UN-VENTILATED."
1675 GET Z$:NO=VAL(Z$):IF NO<>1 AND NO<>2 THEN 1675
1680 PRINT CHR$(147);:REM                         <CLEARS SCREEN!!!!
1685 ON NO GOTO 1695,1820
```

Program listing
Heat loss Contd.

```
1690 REM************* VENTILATED CAVITY SELECTION************
1695 PRINT"MIN WIDTH OF AIRSPACE 0.02M":PRINT
1700 PRINT"1)AIRSPACE BETWEEN ASBESTOS CEMENT OR"
1705 PRINT"  BLACK METAL CLADDING WITH UNSEALED"
1710 PRINT"  JOINTS & HIGH EMISSIVITY LINING."
1715 PRINT"2)AS ABOVE, WITH LOW EMISSIVITY LINING."
1720 PRINT"3)AIRSPACE BETWEEN TILES & ROOFING FELT"
1725 PRINT"  OR BUILDING PAPER ON PITCHED ROOF"
1730 PRINT"4)AIRSPACE BETWN TILE ON TILE HUNG WALL"
1735 PRINT"5)AIRSPACE IN CAVITY WALL CONSTRUCTION"
1740 PRINT"6)LOFT SPACE BETEWEEN FLAT CEILING AND"
1745 PRINT"  UNSEALED ASBESTOS OR BLACK METAL"
1750 PRINT"  CLADDING, PITCHED ROOF."
1755 PRINT"7)AS ABOVE WITH ALUMININIUM CLADDING"
1760 PRINT"  INSTEAD OF BLACK METAL."
1765 PRINT"8)LOFT SPACE BETWEEN FLAT CEILING AND"
1770 PRINT"  UNSEALED TILE ROOF, PITCHED ROOF."
1775 PRINT"9)LOFT SPACE BETWEEN FLAT CEILING AND"
1780 PRINT"  PITCHED ROOF LINED WITH FELT OR"
1785 PRINT"  BUILDING PAPER WITH BEAM FILLING.":PRINT
1790 PRINT"INPUT CHOICE (1-9)"
1795 GET Z$:C1=VAL(Z$):IF C1=0 THEN 1795
1800 PRINT Z$;" SELECTED":FOR X=1 TO TL:NEXT X
1805 R(I,N2)=T3(C1)
1810 GOTO 1610


1815 REM************* UNVENTILATED CAVITY SELECTN************
1820 PRINT"1)WIDTH 0.006M SURFACE EMISSIVITY HIGH."
1825 PRINT"2)WIDTH 0.006M SURFACE EMISSIVITY LOW."
1830 PRINT"3)WIDTH 0.02M  SURFACE EMISSIVITY HIGH."
1835 PRINT"4)WIDTH 0.02M  SURFACE EMISSIVITY LOW."
1840 PRINT"5)HIGH EMISSIVITY PLANE & CORRUGATED"
1845 PRINT"  SHEETS IN CONTACT."
1850 PRINT"6)LOW EMISSIVITY MULT. FOIL INSULATION.":PRINT
1855 PRINT"INPUT CHOICE (1-6)"
1860 GET Z$:C=VAL(Z$):IF C=0 OR C>6 THEN 1860
1865 PRINT:PRINT Z$;" SELECTED":FOR X=1 TO TL:NEXT X
1870 IF D=0 THEN 1885
1875 R1=R1+T4(C,F1)
1880 GOTO 1625
1885 R(I,N2)=T4(C,1)
1890 GOTO 1610


1895 REM************* USER MATERIAL OPTION    ************
1900 PRINT CHR$(147);:REM                    <CLEARS SCREEN!!!!
1905 N=N+1
1910 X=N
1915 PRINT"ENTER DESCRIPTION OF NEW MATERIAL"
1920 PRINT"(MAX 16 CHARACTERS)"
1925 INPUT"DESCIPTION    ";D$(N)
1930 INPUT"K VALUE       ";K(N)
```

Program listing
Heat loss Contd.

```
1935 GOTO 1595
1940 IF D=1 THEN 2105


1945 REM*************    OUTPUT U-VALUE        ***********
1950 N2=N2+1
1955 R(I,N2)=T2(1)
1960 R(I,10)=R(I,10)+1
1965 R=0
1970 FOR J=1 TO N2
1975 R=R+R(I,J)
1980 NEXT J
1985 U(N1,N3)=1/R
1990 PRINT
1995 PRINT"U VALUE FOR THIS CONSTRUCTION=";INT(.5+100*U(N1,N3))/100
2000 IF Q=2 THEN 2025
2005 PRINT
2010 PRINT"ANY MORE U VALUES TO CALCULATE? (Y/N)"
2015 GET Q$:IF Q$<>"Y" AND Q$<>"N" THEN 2015
2020 RETURN
2025 PRINT"ANY MORE TYPES OF CONSTRUCTION"
2030 PRINT"IN WALL NUMBER";N1;"(Y/N)"
2035 GET Q$:IF Q$<>"Y" AND Q$<>"N" THEN 2035
2040 IF Q$="N" THEN 2105
2045 PRINT
2050 L=0:W=0
2055 PRINT"SIZE - LENGTH";
2060 INPUT L
2065 PRINT TAB(6);"BREADTH";
2070 INPUT W
2075 N3=N3+1
2080 IF N3<=5 THEN 2095
2085 PRINT"MAX NUMBER IS 4"
2090 GOTO 2105
2095 A(N1,1)=A(N1,1)-L*W
2100 A(N1,N3)=L*W
2105 RETURN
2110 N1=N1+1
2115 IF N1<5 THEN 1340
2120 PRINT CHR$(147);:REM                        <CLEARS SCREEN!!!!


2125 REM*************  ROOF/CEILING U VALUE    ***********
2130 PRINT"CEILING CONSTRUCTION"
2135 PRINT"  1) ROOM ABOVE"
2140 PRINT"  2) FLAT ROOF ABOVE"
2145 PRINT"  3) PITCHED ROOF ABOVE"
2150 GET Z$:IF Z$<>"1" AND Z$<>"2" AND Z$<>"3" THEN 2150
2155 Y=VAL(Z$):PRINT:PRINT Z$;" SELECTED":PRINT
2160 PRINT"  1) HEAT FLOW UP"
2165 PRINT"  2) HEAT FLOW DOWN"
2170 GET Z$:IF Z$<>"1" AND Z$<>"2" THEN 2170
2175 F1=VAL(Z$):PRINT:PRINT Z$;" SELECTED"
2180 ON Y GOTO 2185,2195,2225
2185 R1=T2(F1+1)*2
```

Program listing
Heat loss Contd.

```
2190 GOTO 2200
2195 R1=T2(4)+T1(5,E)
2200 D=1:N2=1
2205 GOSUB 1495:REM                                    <CALC U-VALUE ROUTINE
2210 PRINT"U VALUE FOR THIS CONSTRUCTION=";INT(.5+100*(1/R1))/100
2215 FOR X=1 TO TL*2:NEXT X:GOTO 2285


2220 REM************* ROOF CONSTRUCTION          ************
2225 PRINT CHR$(147);:REM                               <CLEARS SCREEN!!!!
2230 PRINT"PITCHED ROOF CONSTRUCTION":PRINT
2235 PRINT"  1) CORRUGATED ASBESTOS"
2240 PRINT"  2) CORRUGATED ASBESTOS LINED WITH BDS"
2245 PRINT"  3) CORRUGATED IRON"
2250 PRINT"  4) TILES ON BOARD AND FELT"
2255 PRINT"      WITH PLASTER CEILING"
2260 PRINT"  5) TILES ON BATTENS WITH PLASTER CLNG"
2265 GET Z$:IF Z$<>"1" AND Z$<>"2" AND Z$<>"3" AND Z$<>"4" AND Z$<>"5"THEN2265
2270 P=VAL(Z$):PRINT:PRINT Z$;" SELECTED":FOR X=1 TO TL:NEXT X
2275 U1(1)=T5(P,E)


2280 REM************* FLOOR CONSTRUCTION          ************
2285 PRINT CHR$(147);:REM                               <CLEARS SCREEN!!!!
2290 PRINT"FLOOR CONSTRUCTION":PRINT
2295 PRINT"NO PROVISION FOR SOLID FLOORS":PRINT
2300 PRINT"  1) AIR BRICK ON 1 SIDE ONLY BARE BRDS"
2305 PRINT"  2) AIR BRICK ON 1 SIDE ONLY WITH"
2310 PRINT"     PARQUET, LINOLEUM OR RUBBER"
2315 PRINT"  3) AIR BRICKS ON MORE THAN ONE SIDE"
2320 PRINT"     BARE BOARDS"
2325 PRINT"  4) AIR BRICKS ON MORE THAN ONE SIDE"
2330 PRINT"     WITH PARQUET, LINOLEUM OR RUBBER"
2335 GET Z$:IF Z$<>"1" AND Z$<>"2" AND Z$<>"3" AND Z$<>"4" THEN 2335
2340 P1=VAL(Z$):PRINT:PRINT Z$;" SELECTED":FOR X=1 TO TL:NEXT X
2345 U1(2)=T6(P1)
2350 PRINT CHR$(147);:REM                               <CLEARS SCREEN!!!!
2355 INPUT"ROOM TEMP          (DEG C)";T1
2360 PRINT"OUTSIDE TEMPS      (DEG C)":PRINT
2365 FOR I=1 TO 4
2370 PRINT"                   WALL";I;
2375 INPUT T(I)
2380 NEXT I:PRINT
2385 INPUT"TEMP UNDER FLOOR     (DEG C)";T2
2390 INPUT"TEMP ABOVE ROOF      (DEG C)";T4
2395 INPUT"TEMP OF INCOMING AIR (DEG C)";T5
2400 IF Y<>1 THEN 2410
2405 INPUT"TEMP OF ROOM ABOVE   (DEG C)";T3
2410 INPUT"NO. OF AIR CHANGES PER HOUR ";A


2415 REM************* CALC.STEADY STATE HEATLOSS ************
2420 L0=A*V*134/360*(T1-T5)
2425 FOR I=1 TO 4
2430 FOR J=1 TO 4
```

Program listing
Heat loss Contd.

```
2435 LO=LO+U(I,J)*A(I,J)*(T1-T(I))
2440 NEXT J,I
2445 LO=LO+U1(2)*A1*(T1-T2)
2450 IF Y=1 THEN 2465
2455 LO=LO+U1(1)*A1*(T1-T4)
2460 GOTO 2470
2465 LO=LO+U1(1)*A1*(T1-T3)
2470 PRINT CHR$(147);:REM                        <CLEARS SCREEN!!!!
2475 PRINT"STEADY STATE HEAT LOSS=";INT(LO)/1000;"KW":PRINT
2480 PRINT"ANY MORE CALCULATIONS? (Y/N)"
2485 GET Z$:IF Z$<>"Y" AND Z$<>"N" THEN 2485
2490 IF Z$="N" THEN 9999


2495 REM*************   SETS ARRAYS TO ZERO    ************
2500 REM  IF MACHINE HAS MATH MAT FACILITY LNS 2510-2550 CAN BE REPLACED BY:-
2505 REM   MAT A=ZER: MAT U=ZER :MAT U1=ZER: MAT R=ZER
2510 FOR I=1 TO 4
2515 FOR J=1 TO 4
2520 A(I,J)=0:U(I,J)=0
2525 NEXT J,I
2530 U1(1)=0:U1(2)=0
2535 FOR I=1 TO 16
2540 FOR J=1 TO 10
2545 R(I,J)=0
2550 NEXT J,I
2555 GOTO 1170


2560 REM*************   DATA VALUES FOR TABLE 1 ************
2565 DATA .13,.1,.08,.1,.08,.05,.08,.05,.03,.08,.05,.01,.07,.04,.02


2570 REM*************   DATA VALUES FOR TABLE 2&6************
2575 DATA .12,1.7,.11,1.4,.15,2.3,.11,2


2580 REM*************   DATA VALUES FOR TABLE 3  ************
2585 DATA .16,.3,.12,.12,.18,.14,.25,.11,.18


2590 REM*************   DATA VALUES FOR TABLE 4  ************
2595 DATA .11,.11,.18,.18,.18,.21,.35,1.06,.09,.11,.58,1.76


2600 REM*************   DATA VALUES FOR TABLE 5  ************
2605 DATA6.8,7.9,9.7,2.7,2.8,3,7.1,8.5,10.2
2610 DATA1.6,1.7,1.8,2.8,3.2,3.6,5.7,6.8,7.9


2615 REM*************   MATERIALS AND K VALUES   ************
2620 REM  IN THE FOLLOWING DESCRIPTIONS ABBREVIATIONS HAVE BEEN
2625 REM  USED TO KEEP WITHIN A MAX OF 16 CHARACTERS.IF THE
2630 REM  MACHINE USED HAS THE CAPABILITY OF EXTENDING THESE
2635 REM  AND A WIDE ENOUGH SCREEN FULL DESCIPTIONS CAN BE USED
```

Program listing
Heat loss Contd.

```
2640 DATA"BWK-LIGHT",.806,"BWK-AVERAGE",1.21,"BWK-DENSE",1.47
2645 DATA"CONC-DENSE",1.44,"CONC-CLINK.AGG",.403,"CONC-FORMED SLAG",.245
2650 DATA"PLASTER-GYPSUM",.461,"PLASTER-VERMIC",.201,"RENDER SAND & CT",.532
2655 DATA"PLASTERBD-GYPSUM",.158,"SOFTWOOD",.138,"HARDWOOD",.16
2660 DATA"WOOD FIB SOFT BD",.065,"WOOD WOOL SLAB",.093,"GLASS WOOL QUILT",.034
2665 DATA"EEL GRASS BLKT",.043,"GLASS",1.052,"EXP POLYSTYRENE",.0328
9999 END
```

```
HEAT LOSS CALCULATIONS
========================
THIS PROGRAM IS FOR THE CALCULATION OF
U VALUES FOR VARIOUS CONSTRUCTIONS.

THE PROGRAM WILL CONTINUE IF DESIRED
TO CALCULATE THE STEADY STATE HEAT LOSS
FROM A RECTANGULAR ROOM.

ARE CALCULATIONS REQUIRED OF:-

1) U-VALUES ONLY.

2) U-VALUES & HEATLOSS.

SELECT OPTION REQUIRED.

2 SELECTED.
```

Screen 1

Titles and selection of program routine, i.e. the program to calculate *U*-values only or the program to undertake this and calculate heat loss for the room.

The following displays are for option 2.

```
ROOM DIMENSIONS (M)

LENGTH? 10
WIDTH ? 8
HEIGHT? 3

WALLS 1 & 3 ARE CONSIDERED TO BE
IN THE LENGTH OF THE ROOM.
WALLS 2 & 4 ARE CONSIDERED TO BE
IN THE WIDTH OF THE ROOM.

TOUCH 'RETURN' TO CONTINUE.
```

Screen 2

Input of room dimensions.

```
DETAILS OF WALL CONSTRUCTION

WALL NUMBER 1
IS THIS AN EXTERNAL WALL? (Y/N)

WALL FACING
1) S
2) W, S.W , S.E
3) N.W
4) N, N.E , E

4 SELECTED

CONDITIONS OF EXPOSURE
1) SHELTERED.
2) NORMAL.
3) SEVERE.

2 SELECTED
```

Screen 3

Input of wall position and orientation.

Fig. 11.12 Major screen displays for 'Heat loss'.

Fig. 11.12 Contd.

```
MATERIALS ARE:-
 1)BWK-LIGHT          2)BWK-AVERAGE
 3)BWK-DENSE          4)CONC-DENSE
 5)CONC-CLINK.AGG     6)CONC-FORMED SLAG
 7)PLASTER-GYPSUM     8)PLASTER-VERMIC
 9)RENDER SAND & CT  10)PLASTERBD-GYPSUM
11)SOFTWOOD          12)HARDWOOD
13)WOOD FIB SOFT BD  14)WOOD WOOL SLAB
15)GLASS WOOL QUILT  16)EEL GRASS BLKT
17)GLASS             18)EXP POLYSTYRENE
19)                  20)

MATERIALS TO BE ENTERED STARTING
FROM OUTSIDE ROOM
 MATERIAL NUMBER 1
(CAVITY COUNTS AS A MATERIAL)
IF MATERIAL IS IN LIST ENTER NUMBER
IF NEW MATERIAL IS TO BE ADDED ENTER 50
MATERIAL CODE ? 2
THICKNESS (MM)? 115
```

Screen 4

Input construction of wall.

```
ENTER DESCRIPTION OF NEW MATERIAL
(MAX 16 CHARACTERS)
DESCIPTION      ? PLASTER SPEC
K VALUE         ? 0.215
THICKNESS (MM)? 10
```

Screen 5

Optional input if a material not in the data base is used.

Fig. 11.12 Contd.

```
MIN WIDTH OF AIRSPACE 0.02M

1)AIRSPACE BETWEEN ASBESTOS CEMENT OR
  BLACK METAL CLADDING WITH UNSEALED
  JOINTS & HIGH EMISSIVITY LINING.
2)AS ABOVE, WITH LOW EMISSIVITY LINING.
3)AIRSPACE BETWEEN TILES & ROOFING FELT
  OR BUILDING PAPER ON PITCHED ROOF
4)AIRSPACE BETWN TILE ON TILE HUNG WALL
5)AIRSPACE IN CAVITY WALL CONSTRUCTION
6)LOFT SPACE BETEWEEN FLAT CEILING AND
  UNSEALED ASBESTOS OR BLACK METAL
  CLADDING, PITCHED ROOF.
7)AS ABOVE WITH ALUMININIUM CLADDING
  INSTEAD OF BLACK METAL.
8)LOFT SPACE BETWEEN FLAT CEILING AND
  UNSEALED TILE ROOF, PITCHED ROOF.
9)LOFT SPACE BETWEEN FLAT CEILING AND
  PITCHED ROOF LINED WITH FELT OR
  BUILDING PAPER WITH BEAM FILLING.

INPUT CHOICE (1-9)
5 SELECTED
```

Screen 6

Input airspace characteristics.

```
1)WIDTH 0.006M SURFACE EMISSIVITY HIGH.
2)WIDTH 0.006M SURFACE EMISSIVITY LOW.
3)WIDTH 0.02M  SURFACE EMISSIVITY HIGH.
4)WIDTH 0.02M  SURFACE EMISSIVITY LOW.
5)HIGH EMISSIVITY PLANE & CORRUGATED
  SHEETS IN CONTACT.
6)LOW EMISSIVITY MULT. FOIL INSULATION.

INPUT CHOICE (1-6)

4 SELECTED
```

Screen 7

Input emissivity of surface.

Fig. 11.12 Contd.

```
CEILING CONSTRUCTION
  1) ROOM ABOVE
  2) FLAT ROOF ABOVE
  3) PITCHED ROOF ABOVE

3 SELECTED

  1) HEAT FLOW UP
  2) HEAT FLOW DOWN

1 SELECTED
```

Screen 8

Input type of ceiling/roof to room and direction of heat flow.

```
PITCHED ROOF CONSTRUCTION

  1) CORRUGATED ASBESTOS
  2) CORRUGATED ASBESTOS LINED WITH BDS
  3) CORRUGATED IRON
  4) TILES ON BOARD AND FELT
     WITH PLASTER CEILING
  5) TILES ON BATTENS WITH PLASTER CLNG

4 SELECTED
```

Screen 9

Input type of construction.

```
FLOOR CONSTRUCTION

NO PROVISION FOR SOLID FLOORS

  1) AIR BRICK ON 1 SIDE ONLY BARE BRDS
  2) AIR BRICK ON 1 SIDE ONLY WITH
     PARQUET, LINOLEUM OR RUBBER
  3) AIR BRICKS ON MORE THAN ONE SIDE
     BARE BOARDS
  4) AIR BRICKS ON MORE THAN ONE SIDE
     WITH PARQUET, LINOLEUM OR RUBBER

4 SELECTED
```

Screen 10

Input type of floor construction (hollow floors only).

```
ROOM TEMP            (DEG C)? 20
OUTSIDE TEMPS        (DEG C)

                     WALL 1 ? 0
                     WALL 2 ? 20
                     WALL 3 ? 16
                     WALL 4 ? 0

TEMP UNDER FLOOR     (DEG C)? 15
TEMP ABOVE ROOF      (DEG C)? 0
TEMP OF INCOMING AIR (DEG C)? 0
NO. OF AIR CHANGES PER HOUR ? 3

STEADY STATE HEAT LOSS= 8.817 KW

ANY MORE CALCULATIONS? (Y/N)
```

Screen 11

Input temperature differences and air changes.

Results of steady state heat loss shown.

Daylighting

Introduction

This program is based on the UK *Building Research Establishment Digests* 41 and 42. No account has been taken of the deterioration of the decorations or for dirt on the windows. The average internally reflected component is used and the coefficient *C* is taken from table 1 of *BRE Digest* 42. The values in this table are contained in the DATA in line 45.

Aim and objective

The work involved in producing a set of daylight contours using the BRE daylight protractors is quite considerable. This program uses the formulae on which the protractors are based to produce the daylight factors in a room.

Scope of the program

The room is divided into a 0.5 m grid and the daylight factor based on the CIE sky is calculated for each intersection point of the grid. The origin of the grid is the corner of the room to the left hand side of the window and the first line of values to be produced is on a line parallel to the window wall and at a distance of 0.5 m from it. Values are calculated at the side and back walls of the room.

The program is limited to a rectangular room with a maximum dimension of 10 m and the room can contain one window only. External obstructions, if any, are assumed to be of constant height and for the full width of the window.

The running time for this program on a microcomputer is several minutes as a large number of calculations have to be carried out. The actual time will, of course, vary with the size of room chosen. For this reason the results are displayed on the screen as they are produced. They are displayed again on completion.

The program uses two sub-routines. The sub-routine at line 1000 is for the solution of a triangle given three sides and that at line 1030 calculates the sky component using the formula given in *BRE Digest* 41. The same sub-routine is used to calculate the externally reflected component where there is an obstruction. In this case the result is divided by 10.

Test runs on this program have shown reasonably close agreement with the values found using the BRE daylight protractors. Exact agreement is unlikely owing to the approximations which have to be made when using the protractors.

Machine requirements
- (i) Central processing unit/keyboard
- (ii) Screen
- (iii) Cassette tape deck and blank tape or disk drive and disk for storing the program.

Data requirements
The data which must be supplied by the operator is as follows:

- (i) Dimensions of the room (the wall containing the window is referred to as the length).
- (ii) The dimensions of the window and the cill height.
- (iii) The distance of the window from the origin of the grid.
- (iv) Height of working plane, i.e. the height above floor at which the daylight factors are to be calculated.
- (v) Details of external obstructions (if any). The distance from the window and the height of the top of the obstruction above or below the centre line of the window.
- (vi) The internal reflection factors.

Possible alterations and developments

If your machine is capable of handling mathematical matrices this will reduce the running time quite considerably and the alternative lines are given in the REM statements in the program.

This is an obvious case for the use of a printer to obtain a hard copy of the results. If a plotter is available it is possible to extend the scope of the program to draw a set of daylight contours.

There follow the flow chart (fig. 11.13), the program listing, and the major screen displays (fig. 11.14). for 'Daylighting'.

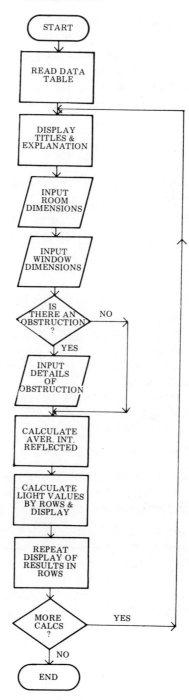

Fig. 11.13 Flow chart for 'Daylighting'.

Program listing
Daylighting

```
100 REM************** DAYLIGHTING STARTER PACK 24-9-81  **************


105 REM**************(C) COPYRIGHT BRANDON,MOORE & MAIN.**************
110 REM*              DIMENSION ARRAYS TO HOLD VALUES OF
115 REM*              C(BRE DIGEST) AND RESULTS
120 DIM C(9),R(21,21)


125 REM**************    READ VALUES INTO C(TABLE 1)    **************
130 FOR I=1 TO 9
135 READ C(I)
140 NEXT I
145 DATA 39,35,31,25,20,14,10,7,5
150 PRINT CHR$(147);:REM                              <CLEARS SCREEN!!!
155 PRINT"THIS PROGRAM IS TO CALCULATE DAYLIGHT"
160 PRINT"FACTORS IN A ROOM HAVING A SINGLE WINDOW"
165 PRINT
170 PRINT"THE WALL CONTAINING THE WINDOW IS"
175 PRINT"CONSIDERED TO BE THE LENGTH":PRINT
180 PRINT"TOUCH 'RETURN' TO CONTINUE"
185 GET X$:IF X$<>CHR$(13) THEN 185


190 REM**************     INPUT ROOM DIMENSIONS     **************
195 PRINT CHR$(147);:REM                              <CLEARS SCREEN!!!!
200 PRINT"INPUT DATA"
205 PRINT
210 PRINT"ROOM DIMENSIONS (IN METRES)"
215 PRINT
220 PRINT"LENGTH OF WALL CONTAINING WINDOW"
225 INPUT"          (MAX 10M) ";L1


230 REM**************     CHECK FOR INCORRECT ENTRY     **************
235 N1=INT(L1*2+1)
240 IF N1>0 AND N1<22 THEN 255
245 PRINT"DIMENSION TOO GREAT-TRY AGAIN"
250 GOTO 225
255 INPUT"WIDTH OF ROOM(MAX 10M) ";W1


260 REM**************     CHECK FOR INCORRECT ENTRY     **************
265 N2=INT(W1*2)
270 IF N2>0 AND N2<22 THEN 290
275 PRINT"DIMENSION TOO GREAT-TRY AGAIN"
280 GOTO 255


285 REM**************     INPUT ROOM SPECIFICATIONS     **************
290 INPUT"HEIGHT OF ROOM(METRES) ";H4
295 PRINT
300 PRINT"WINDOW DIMENSIONS(METRES)"
```

Program listing
Daylighting Contd.

```
305 INPUT"LENGTH OF WINDOW        ";L2
310 INPUT"HEIGHT OF WINDOW        ";H1
315 INPUT"CILL HEIGHT(METRES)     ";H2
320 PRINT
325 PRINT"DISTANCE TO WINDOW FROM"
330 INPUT"          CORNER OF ROOM";L3
335 PRINT
340 INPUT"WORKING PLANE HEIGHT    ";H3


345 REM**************   ROUTINE TO INCLUDE OBSTRUCTION  **************
350 PRINT CHR$(147);:REM                          <CLEARS SCREEN!!!!
355 PRINT"EXTERNAL OBSTRUCTIONS"
360 PRINT
365 PRINT"LIMITATION IS SINGLE OBSTRUCTION"
370 PRINT"OF CONSTANT HEIGHT"
375 PRINT
380 PRINT"IS THERE AN OBSTRUCTION(Y OR N)"
385 GET Q$:IF Q$<>"Y" AND Q$<>"N" THEN 385
390 A1=0
395 D1=0
400 H5=0
405 IF Q$="N" THEN 445
410 PRINT
415 INPUT"DISTANCE TO OBSTRUCTION(METRES)";D1
420 PRINT
425 PRINT"HT.OF OBSTRUCTION(+/- TO LEVEL)"
430 INPUT"             CENTRE OF WINDOW";H5
435 A1=ATN(ABS(H5)/D1)


440 REM**************        SET COEFF C        **************
445 C=39
450 IF A1=0 THEN 490


455 REM******A1 HAS BEEN CALCULATED IN RADIANS CHANGE TO DEGREES******
460 A2=A1/π*180
465 C=5
470 IF A2>=80 THEN 490
475 A3=INT(A2/10)+1
480 C=C(A3)-(A2-10*(A3-1))/10*(C(A3)-C(A3+1))


485 REM**************   INPUT OF U-VALUES   **************
490 PRINT CHR$(147);:REM                          <CLEARS SCREEN!!!!
495 PRINT"INTERNAL REFLECTION FACTORS"
500 PRINT"EXPRESSED AS FRACTIONS"
505 PRINT
510 INPUT"WALLS   ";R1
515 INPUT"FLOOR   ";R2
520 INPUT"CEILING";R3
525 PRINT CHR$(147);:REM                          <CLEARS SCREEN!!!!
530 PRINT"CALCULATING AVERAGE INTERNALLY"
535 PRINT"REFLECTED COMPONENT"
```

Program listing
Daylighting Contd.

```
540 B1=L1*W1
545 B2=(2*L1+2*W1)*H4
550 B3=(L1+2*W1)*H4
555 B4=(L1+2*W1)*(H2+H1/2)
560 R4=(B1*(R2+R3)+B2*R1)/(2*B1+B2)
565 R5=(B1*R2+B4*R1)/(B1+B4)
570 R6=(B1*R3+(B3-B4)*R1)/(B1+B3-B4)
575 D=(.85*L2*H1*(C*R5+5*R6))/((2*B1+B2)*(1-R4))
580 PRINT"AVER INTERNAL REF COMPONENT=";D


585 REM**************      SET RESULTS ARRAY TO D     **************
590 FOR I=1 TO N2
595 FOR J=1 TO N1
600 R(I,J)=D
605 NEXT J
610 NEXT I
615 PRINT"CALCULATING SKY & EXT REF COMPONENTS"
620 PRINT"RESULTS ARE PRINTED IN ROWS"
625 PRINT"PARALLEL TO WINDCW WALL"


630 REM**************      LOOP FOR RESULTS     **************
635 FOR I=1 TO N2
640 PRINT
645 PRINT"ROW NUMBER";I
650 J=0
655 IF Q$="N" THEN 665
660 T3=ATN((H2-H3+H1/2+H5)/(D1+I/2))
665 T4=ATN((H1+H2-H3)/I*2)
670 T2=ATN((H2-H3)/I*2)
675 IF Q$="N" THEN 695
680 T=T4-T3
685 T1=T3-T2
690 O=SGN(T1)
695 J=J+1
700 X=L3-J/2+.5
705 Y=H2-H3
710 GOSUB 960
715 F=F2
720 Y=Y+H1
725 GOSUB 960
730 F1=F2
735 IF Q$="N" THEN 770
740 IF O<>1 THEN 770
745 Y=I/2*TAN(T3)
750 GOSUB 960
755 F3=F2
760 T5=T
765 GOTO 780
770 F3=F
775 T5=T4-T2
780 F4=F1
785 GOSUB 990
790 R(I,J)=R(I,J)+Z
```

Program listing
Daylighting Contd.

```
795 IF Q$="N" THEN 830
800 IF O<>1 THEN 830
805 F3=F
810 F4=F2
815 T5=T1
820 GOSUB 990
825 R(I,J)=R(I,J)+Z/10
830 PRINT INT(.5+100*R(I,J))/100,
835 IF J<N1 THEN 695
840 NEXT I


845 REM**************         END OF RESULT LOOP         **************
850 PRINT CHR$(147);::REM                                <CLEARS SCREEN!!!!


855 REM**************  ROUTINE TO RE-EXAMINE RESULTS   **************
860 PRINT"RESULTS ARE PRINTED IN ROWS"
865 PRINT"PARALLEL TO WINDOW WALL"
870 PRINT"TOUCH 'RETURN' TO RE-EXAMINE RESULTS"
875 FOR I=1 TO N2
880 PRINT
885 PRINT"ROW NUMBER";I
890 FOR J=1 TO N1
895 PRINT INT(.5+100*R(I,J))/100,
900 NEXT J
905 PRINT
910 GET X$:IF X$<>CHR$(13) THEN 910
915 NEXT I
920 PRINT
925 PRINT
930 PRINT"RESULTS COMPLETE"
935 PRINT"ARE FURTHER CALCULATIONS REQUIRED?(Y/N)"
940 GETA$:IF A$<>"Y" AND A$<>"N" THEN 940
945 IF A$="Y" THEN 195
950 GOTO 9999


955 REM************** ROUTINE FOR SOLUTION OF TRIANGLES **************
960 S1=SQR(Y↑2+(L2+X)↑2+(I/2)↑2)
965 S2=SQR(Y↑2+((I/2)↑2+X↑2))
970 S=(S1+S2+L2)/2
975 F2=2*(ATN(SQR((S-S1)*(S-S2)/S/(S-L2))))
980 RETURN


985 REM************** ROUTINE TO CALC SKY COMPONENT **************
990 Z1=ATN(SQR(1/(1/(SIN(F3)*SIN(T5))↑2-1)))
995 Z=3/14/π*(F3-F4*COS(T5))+2/7/π*Z1
1000 Z=Z-1/7/π*(SIN(2*T5)*SIN(F4))
1005 Z=Z*100
1010 IF Z>0 THEN 1020
1015 Z=0
1020 RETURN
9999 END
```

```
INPUT DATA

ROOM DIMENSIONS (IN METRES)

LENGTH OF WALL CONTAINING WINDOW
            (MAX 10M) ? 5
WIDTH OF ROOM(MAX 10M) ? 4
HEIGHT OF ROOM(METRES) ? 3

WINDOW DIMENSIONS(METRES)
LENGTH OF WINDOW        ? 2.5
HEIGHT OF WINDOW        ? 1.5
CILL HEIGHT(METRES)     ? 1

DISTANCE TO WINDOW FROM
        CORNER OF ROOM? 1

WORKING PLANE HEIGHT   ? 0.85
```

Screen 1

Input dimensions of the room, the size and position of the window and the height of the work surface.

```
EXTERNAL OBSTRUCTIONS

LIMITATION IS SINGLE OBSTRUCTION
OF CONSTANT HEIGHT

IS THERE AN OBSTRUCTION(Y OR N)

DISTANCE TO OBSTRUCTION(METRES)? 5

HT.OF OBSTRUCTION(+/- TO LEVEL)
            CENTRE OF WINDOW? 0.25
```

Screen 2

Allows an external obstruction to be placed in front of the window.

```
INTERNAL REFLECTION FACTORS
EXPRESSED AS FRACTIONS

WALLS  ? 0.5
FLOOR  ? 0.1
CEILING? 0.6
```

Screen 3

Asks for reflection factors for the internal surfaces.

Fig. 11.14 Major screen displays for 'Daylighting'.

Fig. 11.14 Contd.

```
CALCULATING AVERAGE INTERNALLY
REFLECTED COMPONENT
AVER INTERNAL REF COMPONENT= .879232604

CALCULATING SKY & EXT REF COMPONENTS
RESULTS ARE PRINTED IN ROWS
PARALLEL TO WINDOW WALL

ROW NUMBER 1
  1.12        4.11        11.77       14.36
 14.03       14.03        14.36       11.77
  4.11        1.12         .88
ROW NUMBER 2
  2.54        4.81         8.1        10.34
 10.96       10.96        10.34        8.1
  4.81        2.54         1.45
ROW NUMBER 3
  2.38        3.58         5.03        6.21
  6.78        6.78         6.21        5.03
  3.58        2.38         1.62
```

Screen 4

Outputs the daylight factors for a 0.5 m grid across the room. Because of the limitations of screen size the results are output in rows commencing next to and parallel to the window wall.

Further reading

General

Bradbeer R., De Bono P. and Laurie P. (1982) *The Computer Book*. BBC Publications.

Chandor A. (1977) *The Penguin Dictionary of Computers*. Penguin Books.

Chandor A. (1981) *The Penguin Dictionary of Microprocessors*. Penguin Books.

Evans C. (1980) *The Mighty Micro*. Coronet Books (Hodder and Stoughton).

Freestone, N. K. (1983) *Databases for Fun and Profit* (Granada).

Fry T. F. (1978) *Beginners' Guide to Computers*. Newnes Butterworth.

PSA/CICA (1980) *Micros in Construction*. Construction Industry Computing Association.

The Construction Industry Computing Association (Cambridge, UK) is a computer users' association dedicated to the provision of independent, unbiased advice and information on the use of computers in the construction industry. Its membership is open to all firms with an interest in this, be they an architectural or quantity surveying practice, an engineer, contractor or software house.

Shelley J. (1981) *Microfutures*. Pitman.

Architecture

Broadbent G. (1973) *Design in Architecture* (Chapter 15). J. Wiley & Sons.

Paterson J. (1980) *Architecture and the Microprocessor*. J. Wiley & Sons.

Reynolds R. A. (1980) *Computer Methods for Architects*. Butterworth.

Quantity surveying

Alvey R. J. (1976) *Computers in Quantity Surveying*. The Macmillan Press.

PSA (1981) *Cost Planning and Computers*. HMSO.
RICS/CICA (1981) *Chartered Quantity Surveyors and the Micro-computer*. Royal Institution of Chartered Surveyors.

Data bases

Dean S. M. (1977) *Fundamentals of Data Base Systems*. The Macmillan Press.
Oliver E. and Chapman R. (1981) *Data Processing*. 5th edition. D. P. Publications.
Paterson J. (1977) *Information Methods for Design & Construction*. J. Wiley & Sons.

Programming

Alcock D. (1979) *Illustrating BASIC*. Cambridge University Press.
James M. (1983) *The Complete Programmer*. Granada Publishing.
Prigmore C. (1982) *30 Hour BASIC*. National Extension College.

Computer hardware

Buchsbaum W. (1980) *Personal Computers Handbook*. Sams.
Jacobowitz H. and Basford L. (1967) *Electronic Computers Made Simple*. Made Simple Books, W. H. Allen.
Ladybird Book (1979) *How it Works . . . the Computer*. Ladybird Books.
Morgan E. (1980) *Microprocessors — A Short Introduction*. Department of Industry.

BASIC instructions

NB Not all of these instructions are available on all microcomputers. Please check your own manual. In many of the instructions brackets are required containing the variable name or number e.g. CHR$(147).

Word	Usage
ABS( )	Returns the absolute value of a number. This is the value of the number without regard to the sign.
AND	A Boolean operator. Results in 1 only if both values are 1.
ASC( )	Returns the ASCII code for a specified character.
ATN( )	Returns the arctangent of the argument.
CALL	To transfer control from BASIC to an assembly language program or sub-routine.
CHR$( )	Returns the string value of the specified ASCII code.
CLEAR	Sets all numeric variables to zero and all strings to nulls. May also clear program text from memory.
CLOSE	Closes a logical file.
CLR	Alternative to clear.
CMD	Used to direct output to a peripheral.
CONT	To continue after a halt.
COS( )	Returns the cosine of the argument.
DATA	Creates a list of values to be assigned by READ statements.
DEF FN	Allows special purpose functions to be defined and used within a program.
DEL	Deletes specified program lines.
DIM	Reserves space in memory for an array or string.
END	Causes a program to halt.
EXP( )	Returns the base of natural logarithms (e) raised to a power.
FOR. . . TO	Starts a loop that repeats a set of instructions until an automatically incremented variable reaches a certain value.
FRE( )	Returns the number of bytes free.
GET	Accepts a single character from the keyboard without echoing it to the screen.

GOSUB	Causes the program to branch to the specified line. When a RETURN statement is encountered the program branches back to the statement immediately following the GOSUB statement.
GOTO	The program branches unconditionally to the specified line.
IF	Provides conditional execution of statements based on a relational expression. Used in conjunction with THEN.
INIT	Immediate mode statement usually used to initialise a disk.
INKEY	Alternative to GET.
INPUT	Receives data input from the keyboard or other input device and assigns the value or values entered to the variable or variables specified.
INT( )	Returns the integer portion of a number.
KEYIN	Alternative to GET.
LEFT$( )	Returns the leftmost characters of a string.
LEN( )	Returns the length of a string argument.
LET	The assignment statement — usually optional.
LIST	Immediate mode statement to display one or more lines of a program.
LOAD	Loads a program into memory from an external device.
LOG( )	Returns the natural logarithm of the number.
MID$( )	Returns any specified portion of a string.
NEW	Deletes the current program from memory.
NEXT	Terminates the loop set up by a FOR instruction.
NOT	A Boolean operator which logically complements each value.
NOTRACE	Turns off the TRACE mode.
ON	Provides for conditional branching. May be used in conjunction with either GOTO or GOSUB.
ONERR	Allows program control of machine discovered errors. May also appear as ON ERROR, ERROR etc.
OPEN	Used to open a logical file.
OR	A Boolean operator. Results in 1 if either value is 1.
PEEK( )	Returns the contents of a memory location.
POKE	Stores a byte of data in a specified memory location.
POS( )	Returns the column position of the cursor. On some machines returns the position of a specified character within a string.
PRINT	Outputs characters to the screen or other output device.

READ	Assigns values from a DATA statement.
REM	Remark statement — allows comments to be placed in a program.
RESTORE	Sets the DATA list pointer to the beginning of the list.
RETURN	Causes the program to branch to the statement immediately following the most recently executed GOSUB.
RIGHT$( )	Returns the rightmost characters of a string.
RND( )	Returns a random number.
RUN	Immediate mode statement which executes the program currently in memory.
SAVE	Saves the program currently in memory on tape or disk.
SGN( )	Determines whether the sign of a number is positive or negative, or if the value is zero.
SIN( )	Returns the sine of an angle.
SPC( )	Moves the cursor to the right a specified number of positions.
SQR( )	Returns the square root of an expression.
SYS	Transfers program control to an independent sub-system.
STEP	Sets the increments to be used in a FOR-TO statement.
STOP	Causes program to halt execution.
STR$( )	Converts a numeric value to a string.
TAB( )	Moves the cursor to the specified position.
TAN( )	Returns the tangent of the angle.
THEN	Part of the conditional statement beginning IF.
TO	Part of the FOR statement.
TRACE	Displays the number of the line being executed. May also display values produced during calculations.
USR( )	Branches to a machine language sub-routine.
VAL( )	Converts a string to a numeric value.
XOR	A Boolean operator. Results in 1 if one and only one value is 1.

Glossary

Acronym — Word formed from the first letter or letters of each word in a phrase, e.g. BASIC, ALGOL.

Address — An identification for a location in storage.

ALGOL — Acronym for ALGOrithmic Language. A universal symbolic language having no reference to the code of a specific computer.

Algorithm — The sequence of logical steps necessary to accomplish a desired goal.

Alphanumeric — Alphabetic and numeric characters.

Architecture — The logical structure of a computing system.

Array — A named, ordered collection of data elements. An array has dimensions specified by a variable defining instruction and its individual elements are referred to by subscripts.

ASCII — A standard code for representing alphabetic information (American Standard Code for Information Interchange).

Assembler — A program used to convert a machine-dependent mnemonic language (assembly language) into the code recognised by the machine (binary).

BASIC — Acronym for Beginners All-purpose Symbolic Instruction Code. An interpreted high level language widely used in microcomputers and small business systems. Easy to understand and simple to employ.

Baud — Unit of speed in telegraphic code transmission. In a binary system the baud rate is the same as the number of bits per second.

Binary — A number system based on the powers of two using the digits 1 and 0.

Bit — An abbreviation of 'binary digit'. The smallest unit of information, having one of two values — on or off, 1 or 0 etc.

Boolean (operations and values) — Operations that treat the values of 0 and 1 as meaning falsity and truth. The Boolean operations most used are AND, OR, NOT and XOR (exclusive OR).

Bootstrap — A short program, usually stored in ROM, used to get the processor to load its operating system into memory. Generally used to start up a disk-based operating system.

Buffer — *See* Data buffer

Bug — Term for an error in a program or system.

Bus — A line or lines used to interconnect computer system components and provide communication paths for addresses, data and control information which need to be transferred between components.

Byte — A set of binary digits considered as a unit, usually a subdivision of a word. Generally accepted as 8 bits and storing one character.

Central processor unit (CPU) — That unit of a computing system which fetches, decodes and executes programmed instructions and maintains the status of results as the program is executed.

Character — One of a set of alphanumeric symbols which may be represented by a unique binary code pattern.

Chip — Popular name for an integrated circuit device.

Clock — A pulse generator which produces basic timing signals to which all system operations are synchronised.

COBOL — Acronym for COmmon Business Oriented Language.

Concatenation — To join together or connect data sets or character strings.

Compiler — A program which takes as input another program, written in a language which people can understand and produces as output another program which is meaningful to the computer.

Computer — A machine for performing computations. Normally operates under a set of programmed instructions.

Configuration — Term to describe the physical units comprising a computer system.

Console — The control device used by computer operators. Usually a teletypewriter and visual display unit, enabling the operator to input messages to the computer and to receive messages back.

Control circuit — A circuit that supervises a particular set of operations.

Conversational mode — A method of operation in which the user is in direct communication with the computer and is able to obtain immediate response to his input messages.

cps — Characters per second.

CRC — Cyclic redundancy check. A form of error checking in which a check pattern is written at the end of each block of data. Commonly used for magnetic tape and disk systems.

Cursor — A means of identifying a point on the surface of a digitiser or a visual display screen.

Data — General term covering the wide range of information on which a computer operates.

Data bank — A collection of data usually stored on magnetic tape or disk and available to users via remote terminals. The information tends to be limited to a particular application area over a particular period of time.

Data base — An organised pool of shareable data — typically a series of regularly updated files related through one or more indices which permit the direct retrieval of information for a wide range of purposes.

Data buffer — A register or small section of memory in the processor or a peripheral device which is used temporarily to hold data when the peripheral device and the CPU are operating at different data rates. In a printer system the buffer may hold a complete line of text or perhaps a complete page.

Data structure — A group of items of information ordered or related in a specific manner.

Debugging — The process of detecting and correcting errors in the operation of a program.

Decimal — A number system based on the powers of 10.

Diagnostic — A special program designed to test a computer system and indicate any faults in operation.

Digitiser — A device to enable information to be taken from a drawing in the form of co-ordinates for input to a computer.

Disk — A magnetically coated disk used for mass storage of data in a computer system. Disks may be of the hard type, such as the Winchester disk, or may be of flexible plastic in a protective envelope, as in the case of the floppy disk.

Disk storage — A method of bulk storage of data and programs. The medium is a rotating circular plate coated with magnetic material. Data is written (stored) and read (retrieved) by fixed or moveable read/write heads positioned over tracks on the surface of the disk. Addressable portions can be selected for read and write operations.

Diskette — An alternative name for a 200 mm diameter floppy disk.

Drive — Used to describe the actual unit in which the magnetic storage media reside.

Execute — To perform a specified computer instruction. To run a program.

Field — Part of a record, used to hold one item of data.

File — An organised collection of records held within the computer system, normally on the intermediate storage device. Such files can be organised in a number of different ways.

Firmware — Software that has been hardwired, i.e. made into a kind of hardware.

Fixed point — A form of arithmetic data storage where data is maintained in binary form with a fixed decimal point.

Floating point — A form of arithmetic data storage where values are maintained in the form of scientific notation.

Floppy disk — A disk made of flexible plastic contained within a paper envelope. The storage capacity ranges from about 80 kbyte to 1 Mbyte.

FORTRAN — Acronym for FORmula TRANslator. A problem oriented high level language designed for scientific and mathematical use.

Function — That part of a computer instruction which specifies the operation to be performed.

Hard copy — A printed output or permanent message, as opposed to the volatile display on a video terminal.

Hardware — The physical units making up a computer system. The apparatus as opposed to the programs. *Contrast with* Software.

Hexadecimal — A number system based on powers of 16 using as digits: 0, 1, 2, 3, 4, 5, 6, 7, 8, 9, A, B, C, D, E and F.

High level language — A language which can be readily understood by a programmer.

High order — The most significant digits in a number or word.

Housekeeping — Standard routines incorporated into a program to perform procedures such as input and output.

Input/output (I/O) — General term for the process of transferring data between the central processor and a peripheral device.

Instruction — A set of bits that defines a computer operation and is a basic command understood by the CPU. It may move data, do arithmetic, perform logic functions, control I/O devices, or make decisions as to which instruction to execute next.

Instruction set — The set of instructions available with a given computer. In general, instruction sets can vary significantly from one computer manufacturer and model to another. The number of instructions only partially indicates the quality of an instruction set. Some instructions may only be slightly different from one another; others may be used only rarely.

Intelligent terminal — A terminal with its own processor and memory. Its function is to carry out simpler tasks locally, using the more valuable time of the computer to which it is linked for tasks beyond the terminal's capability.

Integrated circuit — An electronic circuit formed on a single piece of semiconductor material and then packaged as a plug-in device.

Interactive — A method of operation in which the user is in direct communication with the computer and in which his decisions are influenced by the results obtained.

Interpreter — A program which takes as input one statement of another program written in a language meaningful to people and produces as output a series of instructions which are then carried out by the computer (*compare with* Compiler).

kilobyte — Computer terminology for 1000 bytes or words (actually

2 to the power of 10 = 1024). Symbol kbyte or K.

Language — A means of communicating information between man and machine which can be understood by both.

Loop — A part of a program that is executed repeatedly.

Low level (language) — A language readily understood by the computer. More precisely, simple mnemonic representations of the computer instruction set — called an assembly language. One assembly level statement is equivalent to one machine instruction.

Low order — The least significant digits in a number or word.

Machine code — A sequence of binary machine instructions in a form executable by the computer.

Main frame — A vague term applicable to computer systems which use a large word size and can support large amounts of backing store.

Maintenance

(i) Program maintenance — the upkeep of programs and documentation.

(ii) File maintenance — the processing of a file to bring it up to date.

(iii) Hardware maintenance — the regular scheduled servicing and fitting of manufacturer's modifications and the repair of faults etc.

megabyte — One million bytes. Symbol Mbyte or M.

Memory — The part of the computer which holds the data and instructions. Each instruction or byte of data is assigned a unique address which is used by the CPU when fetching or storing the information.

Memory size — The total number of words (or bytes) that can be stored in a particular memory. Since one location stores one word (or byte), memory size also refers to the total number of locations in the memory.

Menu — A range of options for selection by the user.

Microcomputer — A fully operational computer in its simplest form consisting of CPU and memory. Characterised by its small size, low power consumption and low cost.

Microprocessor — A single package of circuits for the performance of complex logic functions which acts as a CPU.

Minicomputer — A general purpose computer characterised by its relatively low cost, small size and tolerance to changing environmental conditions.

Mini-floppy — A smaller version of the standard floppy disk (130 mm diameter).

Non-volatile memory — A type of computer memory offering preservation of data during power loss or system shutdown. Magnetic media systems are non-volatile.

One's complement — A system for representing negative numbers in binary. The negative of a number is formed by changing all ones in the original number to zeros and all zeros to ones. This is also equivalent to the Boolean NOT operation on each bit.

Operand — The object on which an operation works.

Operating system — A set of software routines whose function is to control the execution sequence of programs running on a computer, supervise the input/output activities of the programs and support the development of new programs through such functions as assembly, compilation, editing and debugging.

Overlaying — A memory management technique in which only parts of a program are resident in memory at one time. The remaining parts are held in secondary storage and loaded into memory only when required.

Parameter — A quantity that is constant in each case considered but that may not have the same constant values in different cases.

Peripherals — Equipment which can be operated under computer control, e.g. printers, disk drives etc.

Plotter — A device for drawing under computer control.

Port — An input output channel between the CPU and external devices such as keyboards, printers etc.

Portability — The facility for software to function on a variety of hardware.

Program — A sequence of instructions to be performed by a computer in processing a given set of data.

Programming language — A means of specifying instructions to a computer.

Random-access memory (RAM) — A computer memory structured so that the time taken to access any data stored in it is the same.

Readability — The characteristic of a program which makes individuals other than the author able to comprehend its function and process.

Real time clock — A timing device used by the computer to derive elapsed time between events and to control processing of time-dependent data.

Record — A set of one or more fields containing related information.

Register — A single word memory location used to hold temporary data during execution of the program.

Response time — The time between the initiation of an operation from a computer terminal and the receipt of the results at the terminal.

ROM -- Read-only memory. Usually built into the hardware. Can be read, or executed if it is a program, but cannot be changed by the user.

Scalar variable -- A variable that can represent only a single data item.

Signed number -- A number with an algebraic sign, plus or minus.

Software -- Computer programs as opposed to hardware or equipment. An expression covering the collection of programs of instructions within the computer which is not an inherent part of the microelectronics. The software may readily be changed and reinput.

Stack -- A series of registers or a section of memory used to hold addresses or data, in particular for holding the return address during sub-routine operations. Operates as a first-in/last-out memory.

Stack pointer -- A register used to hold the address of the current top location of the stack.

Statement -- An instruction line in a high level language such as BASIC, ALGOL etc.

String -- A sequence of alphanumeric data which will be held in memory as a group of character codes but may be dealt with as a single entity in terms of programming.

Sub-routine -- A part of a program not normally complete in itself but used to perform a specific action. It is available for inclusion in other programs.

Teletypewriter -- A machine resembling a typewriter for transmitting messages over an electrical connection. Teletypewriters are sometimes used to type messages into a computer or to receive computer printout.

Terminal -- A device by which data may be input to, or output from, a computer system. Terminals tend to be remote from the computer.

Transportability -- The ability to move software from one type or model of computer to another.

Turnkey -- A combination of software and hardware offered as a complete package to the user.

Two's complement -- A system for representing negative numbers in the binary number system. The negative number is formed by subtracting the number from zero, and ignoring the 'borrow' that propagates off the left end of the register.

Unsigned numbers -- A positive number that cannot have an algebraic sign.

Variable -- A label which gives identity to an item of information and remains assigned to that item even though the item may change its form or value.

VDU — Visual display unit.

Volatile memory — A read/write memory whose contents are irretrievably lost when the operating power is removed. Virtually all types of read/write semiconductor memories are volatile.

Word — A set of bits or characters which is treated as a single logical unit by the computer. Microcomputer suppliers use the word length in bits as an indication of the device.

Write — The process of transferring data into a memory or other device such as a peripheral.

Index